World Geography

by
John Napp

AGS®
American Guidance Service, Inc.
Circle Pines, Minnesota 55014-1796

Editors:
Barbara Kondrchek, M.Ed., Remedial Reading
Ella Westbrook, B.A.

Cover Designer:
Toltzis Communications, Inc.
Sung-Hae "Dinny" Latshaw

Cover Photo:
B. J. Kondrchek

Printed in the United States of America.

ISBN: 0-88671-539-3 (Previously ISBN: 0-86601-677-5)

Order Number: 80510

A 0 9 8 7 6 5 4 3

Contents

Chapter 1

The Shape of the Planet Earth

Introduction

The information you will learn in the first two chapters of this book is important to your understanding of the rest of the book. Many geographic terms used in the study of the various regions of the world are defined for you in Chapters 1 and 2. You may refer back to these first two chapters at any time during the course of your study.

The Globe. The shape, size, and movements of the earth are important things to consider in the study of geography. A globe, the earth in miniature, shows the *roundness* (spherical) shape of the earth. An example of a globe is shown to the right.

The Size of the Earth. The *diameter* (distance through the center of the earth) is approximately 8,000 miles. The *circumference* (distance around the earth at its widest points) is approximately 25,000 miles.

Movements of the Earth: The Axis. The earth *tilts* (leans) on an axis. The *axis* is an imaginary line that runs through the earth from top to bottom at an angle of 23 $^1/_2$ degrees (23 $^1/_2$ °) from north to south. Look at the diagram of the globe. The axis never changes its position; it is always tilted 23 $^1/_2$ degrees.

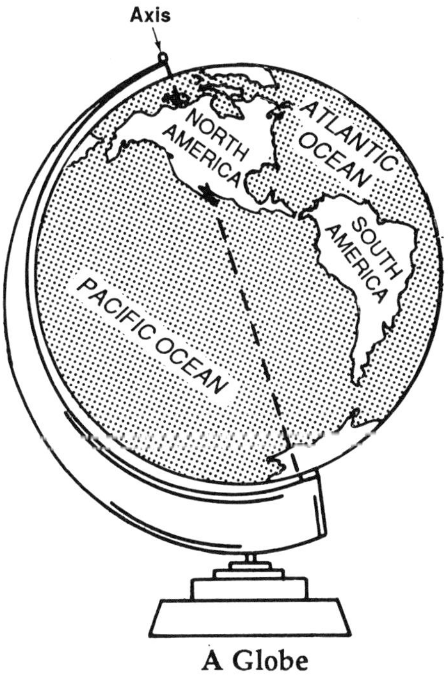

A Globe

Rotation. One way in which the earth moves is called *rotation.* The earth *turns* (spins) on the axis counterclockwise (from west to east). The earth completes one rotation in 24 hours. This is how we complete one day.

Revolution. Another way in which the earth moves is called *revolution.* The earth revolves around the sun. The path or orbit of the earth around the sun takes 365 days. Since one year is really slightly longer than 365 days, the extra quarter of a day is added up, and every four years we have leap year. An extra day is added to February. Notice in the diagram below that the earth is in different positions in relation to the sun at different times of the year. On June 22, for example, the sun's rays are more directly north of the equator. Look closely at each position as the earth revolves around the sun. Seasons of the year change as the earth makes one complete journey around the sun.

Our Earth

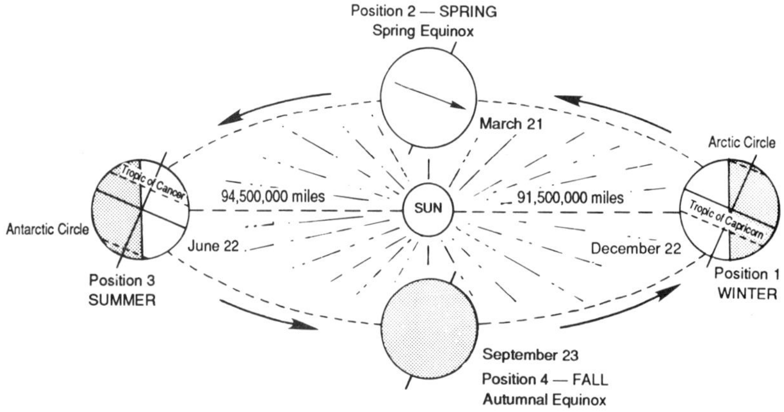

Seasons. In most parts of the world, there are four noticeable seasons. We live in the Northern Hemisphere in the United States; but not all parts of our country experience the great changes that occur from season to season in other parts of our country. For example, in Florida, and other parts of the south, the weather throughout the year is generally rather warm. Parts of California and Arizona also do not have great seasonal changes.

For most of the country, however, changes do take place. As the earth revolves around the sun, the length of the day increases or decreases. When the day is longer, the sun's rays strike the earth more directly. Spring (March 21 to June 21), and summer (June 22 to September 23), are the warmest months. When the length of the day gets shorter, the sun's rays strike the earth at a much less direct angle. Long nights and less warmth from the sun result in much lower temperatures. Fall, sometimes called autumn (September 23 to December 22), and winter (December 22 to March 21) are the coldest months. In the Southern Hemisphere, the exact opposite is true. For example, in Australia, which is south of the equator and in the Southern Hemisphere, winter comes in June, and summer comes in December. Business people and vacationers alike need to know when seasons occur throughout the world.

A. **Directions:** Fill in each blank with the missing term. You will need to use the diagram on page 6.

1. The four important dates for the seasons shown on the diagram on page 6 are ________________, ________________, ________________, and ________________.
2. Because we live in the Northern Hemisphere, our summer begins on ____________.
3. When does our winter begin? ________________________.
4. The two other seasons and dates are ____ and ____.
5. The important tropic line shown in position 3 is the ________________________.
6. The important tropic line shown in position 1 is the________________________.

B. **Directions:** Fill in each blank with the correct term.

round	globe	8,000	night	revolution
25,000	rotation	day	axis	seasons

1. The shape of the earth is ____________________.
2. A____________________ shows the earth in miniature.
3. The diameter of the earth is about ________________________ miles.
4. The circumference of the earth is about ____________________ miles.
5. The earth tilts (leans) on an ____________________________.
6. The movements of the earth are __________________ and ________________.
7. The result of rotation is _______________________ and ______________________.
8. As the earth revolves around the sun, there is a change in the ________________ of the year.

Features of the Earth

The earth's surface is not smooth. It is made up of *landforms* (features) that are called mountains, hills, plateaus, and plains. The highest mountain is Mt. Everest in Nepal, 29,029 feet above sea level. The lowest point is the Dead Sea which is 1,289 feet below sea level. In addition to these features of the earth, there are many rivers. They cut through highlands and find their way to other rivers or to the coastal plains. Rivers are very important because they provide fresh water for home use, as well as for farming and other industries.

The Continents. The earth is made up of much more water than land. Approximately 70 percent of the earth is water and 30 percent is land. The great oceans of the earth, the Pacific, Atlantic, Indian, and Arctic, make up much of that 70 percent. The land area is made up of seven continents. In recent times, six continents are more often identified. Europe and Asia are combined as one and called *Eurasia.* The other five are North America, South America, Africa, Australia, and Antarctica.

Directions: Fill in each blank with the correct term from the box.

Antarctica	Atlantic	six
plateaus	Arctic	water
Pacific	Indian	mountains
70	30	rivers
Africa	Australia	North America
plains	hills	South America

The earth is made up of several landforms such as ____________________, ____________________, ____________________, and ____________________. The surface of the earth is not smooth and has many ____________________ which cut through to the coastal plains. Rivers are very important because they provide a supply of fresh ____________________.

The earth's surface is about ______ percent water and ______ percent land. The great oceans are the ____________________, ____________________, ____________________, and the ____________________.

Lately, ______________ continents are more often identified. They are Eurasia, ____________________, ____________________, ____________________, ____________________, and ____________________.

Latitude. *Latitude* and *longitude* help people understand the earth. These special lines divide the earth into zones. Latitude lines extend around the earth. They are all parallel to one another. The equator is the dividing line between the Northern and Southern Hemispheres. (*Hemi* means half.) *Hemisphere* means "half of the earth." There are 90 degrees of north latitude and 90 degrees of south latitude. Five of the lines of latitude are very special. Of these, the equator is the most important at 0 degrees. The other four are the Tropic of Cancer at 23 $^1/_2$ degrees N., the Tropic of Capricorn at 23 $^1/_2$ degrees S., the Arctic Circle at 66 $^1/_2$ degrees N., and the Antarctic Circle at 66 $^1/_2$ degrees S.

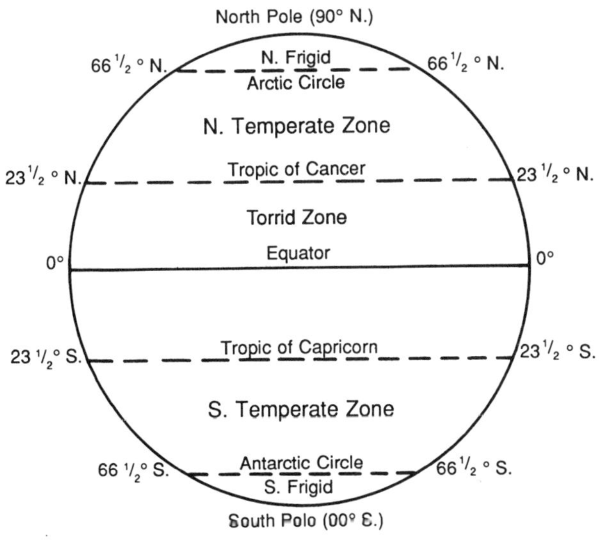

Diagram of Latitude Zones of Light

Zones of Light. The earth is made up of temperature zones. The five major lines of latitude are important in identifying those zones. The equator is the most important of those lines of latitude. The Tropic of Cancer, the Tropic of Capricorn, the Arctic Circle and the Antarctic Circle are also important. The zone between the Tropic of Cancer and the Tropic of Capricorn is called the *Torrid Zone.* The word *torrid* means hot. From the Tropic of Cancer to the Arctic Circle and from the Tropic of Capricorn to the Antarctic Circle is called the *Temperate Zone.* The word *temperate* means moderate (not too hot - not too cold). From the Arctic Circle to the North Pole and from the Antarctic Circle to the South Pole is called the *Frigid Zone.* The word *frigid* means very cold.

The earth has these zones because it tilts on an axis and revolves around the sun. The sun's rays strike different parts of the earth at different angles. The more direct the angle is, the warmer the region. The *indirect* rays (slanted) do not produce as much warmth. Therefore, regions farther from the equator have lower temperatures during the fall and winter seasons. Those regions farthest from the equator have lower temperatures most of the year.

Near the equator, days and nights are generally the same length year-round. In Baltimore, Maryland, the longest day is about 15 hours, and the shortest day is about 9 hours. In Canada, the length of day can vary from 16 hours to 24 hours. The length of night can be 16 to 24 hours. These changes in the length of the days and nights cause a great change in temperatures as well.

A. Directions: Fill in each blank with the missing term. You should refer to the diagram on page 9.

1. The zones of light are the ____________ Zone, ____________ ____________ Zone, ____________ ____________ Zone, ____________ ____________ Zone, and the ____________ ____________ Zone.
2. The Torrid Zone lies between the Tropic of ____________ and the Tropic of ____________.
3. The North Temperate Zone lies between the Tropic of ____________ and the ____________ Circle.
4. The South Temperate Zone lies between the Tropic of ____________ and the ____________ Circle.
5. The North Frigid Zone lies between the ____________ Circle and the ____________ ____________.
6. The South Frigid Zone lies between the ____________ Circle and the ____________ ____________.
7. The most important line of latitude is the ____________.

B. Directions: Match the term in the left column with the correct term in the right column. Write the letter of your answer in the blank.

______	1. latitude	A.	$66^{1}/_{2}$ ° N.
______	2. Arctic Circle	B.	parallels
______	3. Southern Hemisphere	C.	cold
______	4. frigid	D.	0° - 90° S.
______	5. equator	E.	$66^{1}/_{2}$ ° S.
______	6. Tropic of Cancer	F.	$23^{1}/_{2}$ ° N.
______	7. Antarctic Circle	G.	0° - 90° N.
______	8. Northern Hemisphere	H.	$23^{1}/_{2}$ ° S.
______	9. Tropic of Capricorn	I.	0° latitude
______	10. Torrid Zone	J.	hot

Longitude. Longitude lines extend from one pole to the other. Unlike latitude, they are not parallel lines and stop at the poles. As you can see from the diagram below, lines of longitude are not the same distance apart. Only at the equator are longitude lines equal to lines of latitude. The distance of each line of longitude at the equator is about 70 miles. Longitude lines are sometimes called *meridians.* The two most important meridians are the prime meridian, which is at 0 degrees, and the International Date Line, which is at 180 degrees. The prime meridian is especially important because it divides the earth into the Eastern and Western Hemispheres. There are 180 degrees of longitude in the Eastern Hemisphere and 180 degrees of longitude in the Western Hemisphere. We are also north of the equator, so we are in the Northern Hemisphere.

Time Zones. While latitude is important to the change of seasons, longitude is very important in determining time zones. The earth makes one complete rotation every 24 hours. If you divide 360 (the number of degrees for a complete rotation) by 24 (the number of hours to make a complete rotation), each hour of rotation equals 15 degrees. This means that each time zone is about 15 degrees, and there are 24 time zones in the world.

When time zones were established, the prime meridian was placed in Greenwich, England. All time was decided from that point of 0 degrees. Because the earth rotates west to east, places west of the prime meridian have a time which is earlier in the day. For example, if it is noon at the prime meridian, it will be 7:00 a.m. in Philadelphia, Pennsylvania, and 4:00 a.m. in Los Angeles, California. This is 5 hours earlier in Philadelphia and 8 hours earlier in Los Angeles than in Greenwich, England.

Time zones are interesting and very important to understand. If you have relatives in Los Angeles and want to call them, it is helpful to know what time it is there or you could end up calling them in the wee hours of the morning. If your family is going on a trip to England, it would be a good idea to check the time there. The United States (excluding Alaska and Hawaii) has four time zones: Eastern, Central, Mountain, and Pacific.

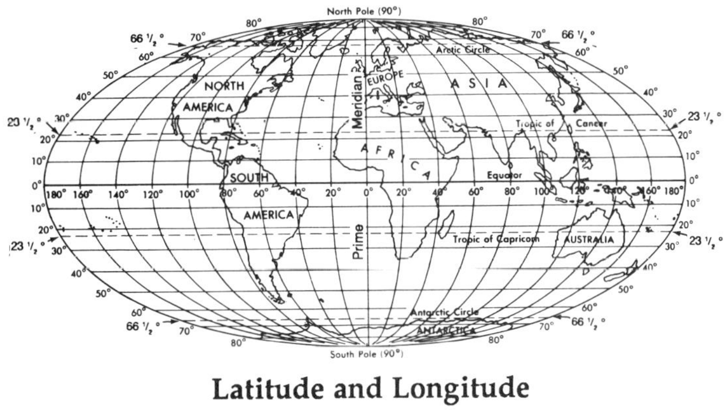

Latitude and Longitude

Grid Lines. Lines of latitude and longitude form what is known as a *grid.* A *grid* is a network of evenly spaced horizontal and perpendicular lines. Latitude and longitude grid lines make it possible to find places on the earth. A similar system is used in your city. Streets are laid out on a grid to make places easier to find.

Directions: Write *True* or *False* for each statement.

_______ 1. Longitude lines are sometimes called meridians.

_______ 2. Longitude lines extend north to south on the earth.

_______ 3. Longitude lines are parallel to one another.

_______ 4. Latitude and longitude are equal in distance only at the equator.

_______ 5. The prime meridian is marked 0 degrees.

_______ 6. All land west of the prime meridian is in the Western Hemisphere.

_______ 7. All land east of the prime meridian is in the Eastern Hemisphere.

_______ 8. There are 180 degrees of west longitude.

_______ 9. The International Date Line is marked the same as the prime meridian, which is at 0 degrees.

_______ 10. Latitude and longitude lines cross and form a grid.

Directions: Complete each of the statements below. One answer can be used twice.

180	prime meridian	15
Greenwich	24	four
5 a.m.	3 p.m.	

1. All time begins at the ________________ ____________________ .
2. The International Date Line is marked ________ degrees.
3. The United States (48 states excluding Hawaii and Alaska) has ______ time zones.
4. Each time zone is approximately ________ degrees.
5. The earth rotates 360 degrees each day and there are ________ time zones.
6. The prime meridian is located in __________________________, England.
7. In order to figure out how many time zones there can be, divide 360 by _______ .
8. If it is 10 a.m. Greenwich time, and you travel five time zones to the west, the time is ________________ .
9. If it is 10 a.m. Greenwich time, and you travel five time zones to the east, the time is ________________ .

Chapter 2
Weather

INTRODUCTION

Weather. In Chapter 1, a variety of topics were presented that described the earth. In Chapter 2, the importance of each can be seen as it relates to weather. A great amount of time, money and interest is given to the weather every day. To a small child, it may only mean that he or she cannot play outside because of the rain. To many others, however, the weather is far more important. Farmers, airline pilots, fishermen, and business people are just some of the many people affected greatly by the weather. At times, it can be so severe, it becomes a danger to people's safety. Weather, therefore, is a subject worth further study.

The Elements of Weather. The condition of the atmosphere produces the daily weather. Important things to consider are: *air pressure, humidity, temperature, wind velocity* and *direction, precipitation, clouds,* and *sunshine*. A weatherman considers all of these items many times a day in order to give accurate weather reports. What exactly does each of these mean? You will study about these items in this chapter.

The Atmosphere. Weather means "the daily condition of the atmosphere at a particular time." The atmosphere is like a blanket that covers the earth. The blanket is in the form of gases, so it is not something we can see. Even though we cannot see the atmosphere, it is possible to examine it and learn a great deal about it.

Air Pressure. Air has weight. A square inch of air at sea level weighs 14.7 pounds per square inch (psi). We do not feel that pressure because we are able to adjust to it. Sometimes, people have problems if they go to higher places where air pressure changes or experience sudden changes in air pressure on airplanes.

A Barometer. The instrument that measures air pressure is called a *barometer*. The weight of air supports a column of mercury, approximately 30 inches at sea level. As the weight of air increases or decreases, the column of mercury goes up or down. The barometer records the changes that occur. Because air is affected by temperature, changes go on throughout the 24 hours of each day and night. While changes are not great, they indicate what the condition of the atmosphere is at a given time. If the barometer reads over 30 inches, it is called *high pressure.* If it reads below 30 inches, it is called *low pressure.* Remember, those readings are at sea level. If you live in higher elevations, normal barometric pressure will generally be one inch lower for every 1,000 feet.

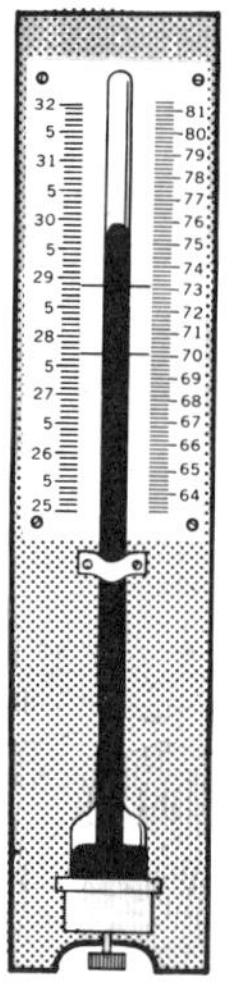

A Barometer

Directions: Complete the exercise.

1. List four groups of people who have special interest in the weather.

   ____________________ ____________________

   ____________________ ____________________

2. Why is weather a very important subject to many people?____________________

   ____

3. Write the definition of weather.____________________

   ____

4. How would you describe the atmosphere of the earth?____________________

   ____

5. List the seven elements (items) that are important in a weather forecast (report).

   a. ____________________ e. ____________________

   b. ____________________ f. ____________________

   c. ____________________ g. ____________________

   d. ____________________

6. How much does air weigh at sea level?____________________
7. Why don't we feel the pressure of air?____________________
8. What instrument is used to measure air pressure?____________________

   How does it work?____________________

   ____

9. If the barometer is over 30 inches at sea level, air pressure is said to be ________ .
10. If the barometer is under 30 inches at sea level, air pressure is said to be ________ .
11. If you live above sea level at 5,000 feet, normal air pressure will be (lower, higher) than 30 inches. (Circle your answer.)

HUMIDITY

Humidity refers to water vapor in the atmosphere. Because water vapor is a gas, we cannot see it. Where does it come from and how does it become a gas? You learned in Chapter 1 that the earth is made up of over 70 percent water. Large oceans, lakes, rivers, and land itself all contribute to the process of producing water vapor. That process is called *evaporation*. We have all seen water disappear. It goes into the atmosphere. The major source of water vapor is, of course, the very large oceans. To measure water vapor in the air, the term *relative humidity* is used. The person giving the weather report says the relative humidity is 60 percent, 20 percent, or whatever it is at that time and at that place. These numbers are in relation to 100 percent. In other words, the air could hold 40 or 80 percent more water vapor at those temperatures. If the atmosphere could not hold any more water vapor, it would be 100 percent or *saturated*. Saturation means "filled to capacity."

Changes in Relative Humidity. As air temperature changes, it can either hold more water vapor or less water vapor. Warmer air can hold more water vapor than colder air. For example, air at 80 degrees Fahrenheit can hold more water vapor than air at 40 degrees Fahrenheit. As temperature increases, relative humidity decreases. As temperature decreases, relative humidity often increases.

Try this example. The relative humidity at 8 a.m. is 70 percent; the temperature is 60 degrees. By 1 p.m. the temperature has gone up to 85 degrees. The higher temperature allows air to hold more water vapor. Would the relative humidity be higher or lower? If you answered lower, you are correct. As the temperature goes up, relative humidity is likely to go down.

A. Directions: Underline the correct answer.

1. Water vapor is a (liquid, gas).
2. Most of the earth's surface is made up of (land, water).
3. Evaporation changes a (liquid to a gas, gas to a liquid).
4. The major source of water vapor is from (rivers, oceans).
5. Water vapor in the air is called (humidity, pressure).
6. The term relative humidity is water vapor in the air compared to (100 percent, 0 percent).
7. When air has all the water vapor it can hold at that temperature, it is said to be (wet, saturated).

(Continued on page 16.)

8. Relative humidity changes many times a day because (temperature, pressure) changes.
9. Warm air can hold (more, less) water vapor than colder air.
10. Generally the rule is—as the temperature goes up, relative humidity is likely to go (up, down).

B. Directions: Define each term.

1. Weather____

____

2. Atmosphere____

____

3. Air pressure____

____

4. Barometer____

____

5. High pressure ____

____

6. Low pressure____

____

7. Humidity____

____

8. Water vapor____

____

9. Saturated____

____

10. Relative humidity____

____

Temperature

In a weather report, people are very interested in hearing what the temperature is or will be for the day or night. Temperature is the warmth or coldness of the atmosphere. The instrument that measures temperature is called a *thermometer*. The word *therm* means heat; a *meter* is an instrument of measurement. Therefore, a thermometer measures temperature.

Fahrenheit Scale. The Fahrenheit scale was introduced by Gabriel Fahrenheit. The scale he developed showed that water boils at 212 degrees and freezes at 32 degrees. The thermometers that most people use only go up to 130 degrees since the atmosphere is not going to be anywhere near 212 degrees. Thermometers also show lower temperatures, 0 degrees and below 0 degrees. In some parts of our country, there is very cold weather. The temperature can go well below 0 degrees.

Centigrade or Celsius Scales. In all other countries except the United States, the Celsius, or centigrade scale is used, in which 100 degrees represents the boiling point of water, and 0 degrees represents the freezing point of water. Very often, both the Fahrenheit and Celsius temperatures are given in a weather report. If you were to visit Canada, our neighbor to the north, and hear a weather report, the temperature could be 10 degrees (°C). On a Fahrenheit scale it would be 50 degrees (°F).

The Importance of Temperature. Temperature plays a very important role in our daily weather. The nature of the atmosphere, which is made up of gases, is to expand when heated and contract when cooled. *Expand* means to take more space; *contract* means to take less space. As the air is warmed, it expands and *ascends* (rises). As the air cools, it contracts and *descends* (comes down). This causes a movement of air upward and downward, often referred to as *convection*. The same idea is used in your house. Warm air from the furnace is blown into a room where it rises and spreads its warmth. The cool air comes down and enters the room's return vent. This process goes on over and over again. Convection in the atmosphere, however, is more complicated and many more things can happen.

Temperature and Wind. All of the elements of weather affect each other in some way. Temperature is very important to all of them. Changes in air temperature bring about changes in air pressure. Cold temperatures over a large region produce a large mass of air which is very heavy and is called *high pressure*. The barometer will show a reading of over 30 inches (sea level). Meanwhile, warm to hot temperatures over other large regions will produce a large mass of air which is much lighter and is called *low pressure*. The barometer will show a reading of under 30 inches (sea level). As the warmer air rises (ascends), and the cooler air drops (descends), the motion of air is called *wind*. The direction of the wind is very important. Also, the speed of the wind is measured. The *speed* or *velocity* of the wind is usually not a problem. At times, however, winds can reach fast speeds and cause great damage.

Directions: Complete the exercise below. Use the terms in the word box. One term may be used twice.

thermometer	warmth	coolness	0	Celsius
Fahrenheit	212	100	contracts	30
32	centigrade	expands	convection	wind

Temperature is the ________________ and ________________ of the atmosphere. The instrument that measures temperature is called a ______________________. The measurement of temperature used in the United States is made on the __________________________ scale. The boiling point of water is __________ degrees, and the freezing point is __________ degrees . Another measurement used in all other countries is the ____________________ or ____________________ scale. The boiling point of water on that scale is ________ degrees, and the freezing point of water is __________ degrees.

When air is heated, it ______________; when air is cooled, it ________________. This causes the upward and downward movement of air which is called ____________.

A high pressure air mass shows a reading on a barometer of over ________ inches. A low pressure air mass shows a reading of under ________ inches on a barometer. The difference in air pressure causes air to have motion called ________________.

PRECIPITATION

Without water, life on earth as we know it to be would not be possible. So much depends on a good supply of water. Much of the earth's surface is made up of large bodies of water. These large oceans and seas, however, are salt water and not suitable for use by people. They are also not suitable for farming. The salt could be removed, but that raises some questions about cost and environmental problems. Where does our water supply come from?

Condensation. Earlier in the chapter, humidity was explained. *Water* (a liquid) evaporates into the atmosphere as a gas. *Evaporation,* therefore, is the changing of a liquid to a gas. The upward movement of air reaches a point where the upper air is much cooler than the surface air. At that point, *water vapor* (gas) forms into droplets of water. That process is called *condensation.* There are many examples of condensation in our daily lives. If you fill a glass with ice cubes and add a cold beverage or water, the outside of the glass gets wet. The cold surface of the glass caused the water vapor to condense. On a much larger scale is the process called the *water cycle.*

The Water Cycle. There are several sources of water vapor on the surface of the earth. Water vapor comes from oceans, lakes, rivers, trees, and from the land itself. Notice the water table in the diagram below. That is where ground water is stored and eventually comes to the surface.

How fast or how slowly the water cycle works depends largely on temperature. Regardless of high temperatures or low temperatures, the cycle still continues to operate.

The evaporation process changes water into a gas and becomes a part of the atmosphere. As the sun's rays strike the earth, *warmth* (heat) is produced. The warmed air rises to a higher point where condensation can take place. The water vapor becomes tiny droplets of water. The tiny droplets of water collect into clouds. As the condensation continues, clouds get larger and droplets of water increase in size. If everything goes right, precipitation can occur. Very often all the parts of the water cycle work as they should, but no precipitation takes place. The supply of fresh water from precipitation is important to everyone.

The water cycle is shown in the diagram below.

The Water Cycle

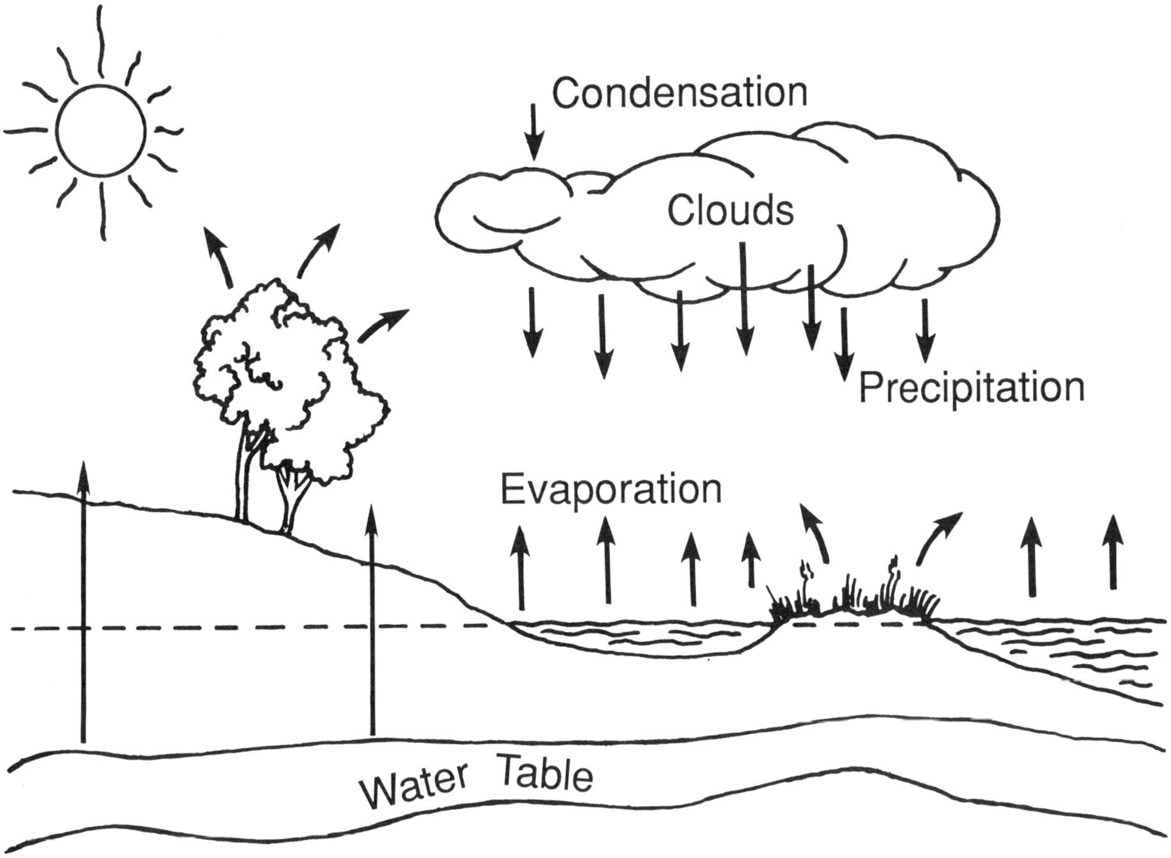

A. Directions: Identify the parts of the water cycle.

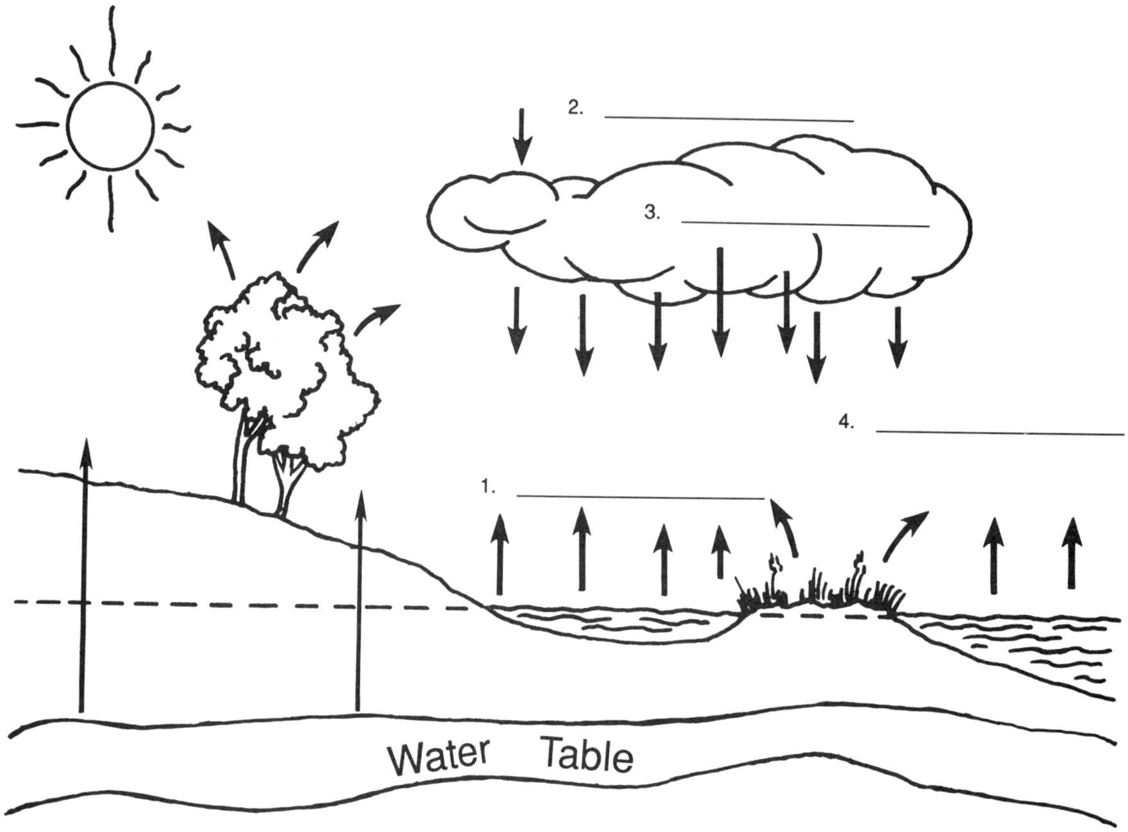

B. Directions: Complete the exercise.

1. List five possible sources of water vapor.

a. ______ c. ______ e. ______

b. ______ d. ______

2. Evaporation changes water from a liquid to a ______.
3. Condensation changes a gas to a ______.
4. Clouds are made up of droplets of ______.
5. The water cycle is the only natural way of producing fresh water that we call ______.

Precipitation. Before precipitation can take place, clouds must form. Depending upon the temperature of the upper atmosphere, rain, snow, sleet, or hail may take place. Rain and snow are two very important forms of precipitation. To a lesser extent, dew, frost and fog are also considered forms of precipitation.

Types of Precipitation

Convectional. Precipitation during warm to hot months of the year often can be very heavy. Thunderstorms develop quickly and generally stop after an hour or so. Sometimes lightning and high winds cause damage to property. Flooding can also take place. These periods of precipitation are called *convectional rains.*

Cyclonic Precipitation. Regions of the middle latitudes have cooler temperatures in the fall and winter months. Rain, sleet, and snow may occur. Precipitation can take place over a large area and very often lasts for hours, even days. The quantity of precipitation, however, will be very small considering the long period. A 12-inch snowfall is only about 1 inch of rain. Days of light rain may only amount to only one-quarter inch to one-half inch of actual precipitation. This type of precipitation is called *cyclonic.*

Directions: Complete the exercise.

1. List two types of precipitation that can occur.
   a. ________________________________
   b. ________________________________
2. During warm months of the year, ____________________ develop quickly.
3. Where does cyclonic precipitation usually take place?

   ____
4. List three facts about cyclonic precipitation.
   a. ________________________________
   b. ________________________________
   c. ________________________________

Clouds and Sunshine

Clouds. Also included in a daily weather report is some mention of clouds and sunshine. Clouds are a part of the water cycle and important for possible precipitation. However, in a weather report, a cloudy day is a way of describing how much or how little sunshine will be seen.

Sunshine. Nothing is as important as sunshine as far as weather is concerned. Sunshine, or light from the sun, provides the earth with its source of energy. It is necessary for life itself. The *warmth* or *coolness* (temperature) of the atmosphere depends upon sunshine. The earth is heated by the sun, and in turn, the earth gives off that heat to the atmosphere.

Three important terms are insolation, radiation, and convection. *Insolation* is the process by which the earth *absorbs* (gets) heat. *Radiation* is the process by which the earth *transfers* (gives off) that heat. The same process applies to coldness. As the air is warmed, an upward and downward movement takes place which is called *convection.*

In a weather report, some mention is made of how much sunshine is likely for that day. Mostly sunny, or partly sunny are ways of describing the day in general terms. People like to know if its going to be a sunny day.

The Process of the Atmosphere Being Heated

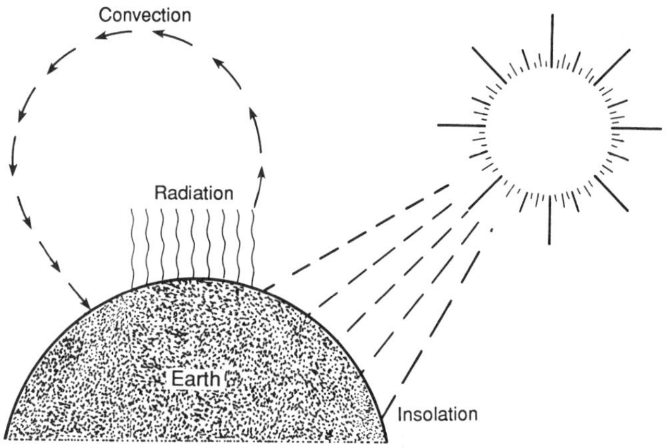

Directions: Complete the exercise.

1. List all the parts (elements) of weather.

   a. ______________________ e. ______________________

   b. ______________________ f. ______________________

   c. ______________________ g. ______________________

   d. ______________________

2. Write your own weather report. Try to use as many of the seven elements of weather as you can. Use your own paper.

3. Define these terms.

   a. Insolation ______________________

   ____

   b. Radiation ______________________

   ____

   c. Convection ______________________

   ____

Chapter 3
Australia

Introduction

Countries are divided into special parts. These are often called *political divisions*. In our country, there are 50 states plus the District of Columbia. In Australia, there are six states and two territories. Not all countries use states. Canada for instance uses *provinces* as its political divisions. Regardless of what they are called, the meaning is still the same. Australia prefers to use states.

In 1901, the six Australian divisions, or colonies, were joined together as states into the country of Australia. The six states are New South Wales, Victoria, Queensland, Western Australia, Tasmania, and South Australia. The Northern Territory and the Australian Capital Territory are the two other divisions of Australia. A *territory* is a region which has too few people to become a state.

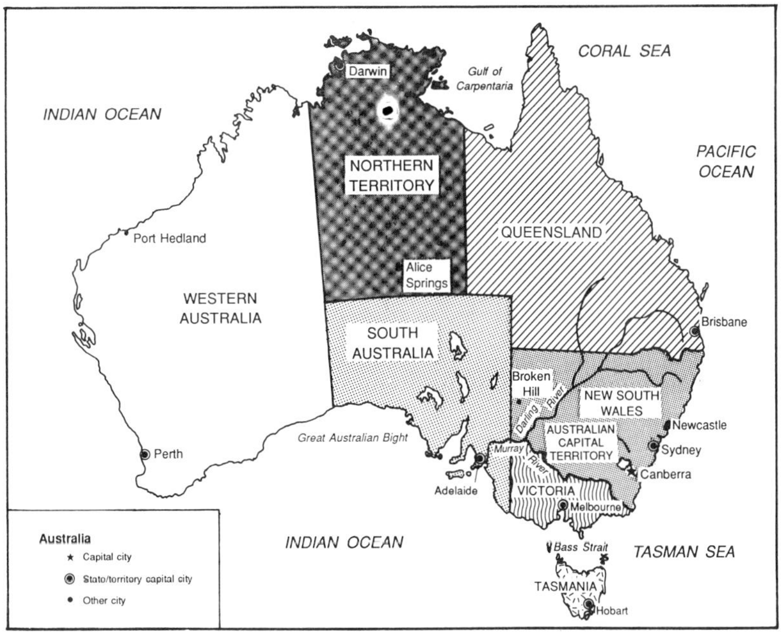

Australia

This chart shows the size (area) and population of the six states of Australia and the two territories.

State & Territory	Area (sq. miles)	Population (1984)
Tasmania	26,178	437,300
Victoria	87,854	4,075,900
Queensland	666,699	2,505,100
South Australia	379,824	1,353,000
New South Wales	309,418	5,405,100
Western Australia	974,843	1,382,600
Northern Territory	519,633	138,900
Australian Capital Territory	926	245,600
TOTALS	2,965,375	15,543,500

*Note: The estimated population in 1987 was 15,700,000.

Directions: Use information from the chart to fill in each blank.

1. Australia's largest state in area is ______________________.
2. Australia's smallest state in area is ______________________.
3. The state of ______________________________ has the largest population.
4. The state of ______________________ has the smallest population.
5. How large in area is the Australian Capital Territory? ________________.

AUSTRALIA

Australia—The Aborigines. Australia was settled by the British in 1788. Englishmen found a native population called *Aborigines,* or *bushmen,* living in Australia. The brown-skinned people were wanderers who fished and hunted to survive. Over the years, the number of Aborigines became smaller. Most of them now live in the Northern Territory. Land has been set aside for the native population much like the reservations for the American Indians of our country.

Industries of Australia. In earlier times, little attention was paid to Australia. The population was very small. People lived in the slow growing cities along the coast. Even as late as 1966, there were only eleven million people in Australia.

Australia is very far from the major industrial countries of the world. Therefore, Australians took care of themselves (self-sufficient). The *Aussies,* as they are sometimes called, produced wool from sheep as well as wheat. Both products became very important to the *economy* (business) of the country. Sugar cane, dairy products, and cattle were also important products that Australians needed and could sell. Slowly, manufactured products increased. Steel mills, automobile assembly plants, clothing factories, flour mills, and sugar refineries are a few of the new industries in Australia. Mining is also a major industry. Minerals such as iron ore, lead, gold, copper, coal, oil, and *bauxite* (aluminum) are very important to Australia and other industrialized countries around the world. The mineral-rich country may have even more minerals yet to be found. No longer is Australia just thought of as a faraway country. In a short time, Australia has become a leading industrial nation.

Population—Urban and Rural. Australia is a land of *city* (urban) people. Nearly 85 percent of the Australians live in cities and towns. The remaining 15 percent are scattered over the other parts of Australia. The outback, or grazing land, of Australia has large *stations* (ranches). Sugar cane plantations and large wheat farms are also a part of the *rural* (non-urban) part of Australia. Even those parts of Australia are beginning to see more towns and small cities spring up as living conditions improve.

Directions: Fill in each blank with the correct term. Choose terms from the box.

85	iron ore	Aussies
flour mills	bushmen	automobile assembly plants
Aborigines	lead	gold
steel mills	city	coal
bauxite	yourself	oil
little	copper	non-urban

1. The native (original) people of Australia are the ______________ .
2. Sometimes the native Australians are called the ______________ .
3. In earlier times, __________ attention was paid to Australia because the country is so far from the major industrial parts of the world.
4. To be self-sufficient means to take care of __________ .
5. Australians are sometimes called ______________ .
6. Rural means ______________ .
7. Some newer industries developed in recent years are __________ __________ , __________ __________ , and __________ __________ .

(Continued on page 26.)

8. Minerals are also important to Australia. The country has large deposits (supplies) of ________ ________, ________, ________, ________, ________, ________, and ________.
9. Urban means ________.
10. About ________ percent of the Australians live in cities.

Major Cities of Australia. Even though Australia's population is rather small, only 15,700,000, some of the cities are very big.

The chart below, will help you see more clearly the important cities of Australia and the state in which each is located.

State & Territory	Represents State Capital	Population
Tasmania	Hobart	150,000
Victoria	Melbourne	2,890,700
South Australia	Adelaide	979,600
Queensland	Brisbane	1,146,600
New South Wales	Sydney	3,358,000
Western Australia	Perth	983,400
Northern Territory	Darwin	65,000
Australian Capital Territory	*Canberra	264,400

*The capital of Australia is Canberra. Canberra is in the Australian Capital Territory.

Directions: Complete the exercise.

1. List the six states of Australia and the state capital of each.

State	Capital
a. ________	________
b. ________	________
c. ________	________
d. ________	________
e. ________	________
f. ________	________

2. List the two largest cities of Australia and the state of each.

City	State
________	________
________	________

Chapter 4
Latin America

INTRODUCTION

Latin America is a part of the earth that stretches from the southern boundary of the United States to the southern tip of South America. The Rio Grande River is the southern boundary of the state of Texas. From that point southward, Latin America begins. As you can see from the map below, Latin America is made up of many countries. Mexico, countries of Central America, island countries in the Caribbean Sea, and countries of South America all make up Latin America. The name Latin America is used because Spanish, Portuguese, French, and Italian explorers went to the region. The language of those explorers had developed from Latin. Latin was the language of an earlier people called *Romans.*

Latin America

The History of Latin America. Early explorers were searching for a way to sail to the *Far East* (Orient). Christopher Columbus made three voyages and thought that he had found the Orient. Instead, he had only discovered the islands in the Caribbean Sea. He never learned that he had been mistaken. Others also made voyages, and soon Latin America became a place of gold and land. Spanish explorers such as Balboa, Pizarro, and Cortes made trips to this new land. The native Indians were soon defeated by strong Spanish armies. Spain established a very big empire in Latin America. Portuguese explorers claimed what is now Brazil for Portugal. The colonies that had been established gradually became independent, but the changes were not easy to make. There are still some major political problems in Latin America.

Directions: Use the information you have read to complete the exercise.

1. Latin America begins at the ________________ ________________ River and extends to the tip of ____________________ ______________________ .
2. Latin America can be divided up into four parts. They are the:
   a. country of ______________________________ .
   b. islands in the __________________ ____________________.
   c. countries of __________________________ ___________________________ .
   d. continent of __________________________ ___________________________ .
3. Latin America gets its name from a language spoken by the early ______________.
4. List four early explorers who went to Latin America.
   a. ______________________________ c. ________________________________
   b. ______________________________ d. ________________________________
5. Which European country claimed Brazil? ____________________________________.
6. Who are the natives of Latin America? ___________________________________ .
7. Gradually, the European colonies became ___________________________________ .

Land Features of South America. Latin America is a region of great differences in *land features* (topography). The Andes Mountains of western South America cover a wide range of latitudes. The *spine* or backbone of the continent has peaks 20,000 feet above sea level. The Andes Mountains extend from 10 degrees N. latitude all the way to 55 degrees S. latitude. The Andes Mountains can be found in the western countries of Venezuela, Colombia, Ecuador, Peru, Bolivia, Chile, and western Argentina. The lower lands are located in the eastern part of South America. The two most important are the *Pampas* (grasslands) of Argentina and the *Amazonian Basin* (lowlands) of Brazil.

Land Features of Central America. Central America is called the land bridge between North America and South America. The mountains extend throughout the region. Some peaks are 10,000 feet above sea level. Deep valleys, coastal plains, and plateaus are also important land features in the seven small countries.

Land Features of Mexico. Mexico is mostly *plateau* (table top), flat lands of 4,000-8,000 feet. However, very high mountains can be found. Some mountains are 15,000 feet above sea level.

The islands of the West Indies in the Caribbean Sea are also made up of highlands. Scattered among the islands are coastal plains and gently sloping land suitable for farming.

Latin America, as you can see, is quite a mixture of very high lands, plateaus, valleys, and lowlands. The *topography* (features) influences the climates of the region.

Directions: Use the information you have read to complete the exercise.

1. What is another word for land features?____________________
2. The Andes Mountains are located in the western part of ____________________ .
3. The Andes Mountains are called the spine, or backbone, of South America because they extend from _______________ N. latitude to _______________ S. latitude.
4. The Andes Mountains can be found in the following South American countries:

   a. ____________________ e. ____________________

   b. ____________________ f. ____________________

   c. ____________________ g. ____________________

   d. ____________________
5. List the two major lowland regions of South America.

   a. ____________________ b. ____________________
6. Central America is sometimes called the _______________ _______________ between North America and _______________ _______________ .
7. The land features of Central America are _______________ , _______________ , _______________ _______________ , and _______________ .
8. How many Central American countries are there? _______.
9. Mexico is mostly made up of a large _______________. There are, however, some _______________ 15,000 feet above sea level.
10. Land features (topography) greatly influence _______________ .

Mexico

Mexico and the United States. There are many countries in Latin America. One of those countries borders our country to the south. That country is Mexico. The boundary line between the United States and Mexico is the Rio Grande River. The boundary line was *established* (set) in 1848 when the war with Mexico ended. Before the war with Mexico in 1846, much of the southwestern part of the United States had been a part of Mexico. California and New Mexico are both *Spanish* (Mexican) names.

The Geography of Mexico. Mexico is a large country of 761,000 square miles in size with a population of about 82,000,000. The Pacific Ocean lies to the west, and the Gulf of Mexico lies to the east. The United States borders Mexico to the north, and Central America borders Mexico to the south.

Mexico is very much like our western states. There are mountains, plateaus, and a coastal plain in Mexico. Most of the land, however, is a plateau that is 4,000 to 8,000 feet high. Scattered in between are the high mountains which are 15,000 feet and more above sea level.

With such a great difference in topography, the climates of Mexico are quite different from one part to another. Some parts are very dry; other parts are very wet. Some parts are very hot, with a temperature of 100 degrees, and at times even hotter. Higher lands are often much cooler. Mexico City, the capital city, is 7,487 feet above sea level. Mexico City also has a very large population because the mild pleasant climate is very comfortable.

Mexico Today. For many years, Mexico has been a country which depended very much on farming as its major industry. Farms are small and require hard work. The larger farms, called *haciendas,* do much better but are owned by richer people. The Mexican workers on those large farms do not earn high wages.

Some new industries have begun in Mexico. Oil and other minerals are now being developed, but progress has been slow. Many Mexicans are unskilled and uneducated, which only makes the problems worse. The government does not have the money to do much about all the problems that Mexico faces. The United States has made money available in the form of loans to Mexico. However, it will take time for Mexico to raise the standard of living of its people and become an industrial nation.

Directions: Define these terms.

1. spine ____

2. Pampas ____

(Continued on page 31.)

3. plateau ____

4. hacienda ____

5. topography ____

Brazil

The Geography of Brazil. The largest country of South America is Brazil. Its 3,286,400 square miles cover nearly half of the entire continent of South America. Only the Soviet Union, China, Canada, and the United States are larger in area than Brazil. The population is also the largest in South America. More than half of all the people of South America live there. With a population of about 143,300,000 people, and over 3 million square miles of land, you would think that Brazil was not very crowded. But once again, land features and climates play an important role in deciding where people can live and work.

Land Features of Brazil. The lowlands of the Amazon River Basin make up a very large part of Brazil. The river flows some 3,900 miles across Brazil and is very near to the equator. The Amazon River is not as long as the Nile River of Africa but much bigger in terms of the quantity of water it carries.

A plateau rises 3,000 feet in the southeastern part of Brazil. The land is high enough to provide for cooler temperatures.

Between the narrow coastal plains and the plateau region lies the Brazilian Highland. The Highlands rise steeply from the coastal plain. The mountain range is not very high, except for those near the city of Rio de Janeiro. West of the city, the mountains reach 8,000-9,000 feet above sea level.

Brazil

Directions: Fill in each blank with a term from the box.

Brazil	United States	Canada
plateau	China	Soviet Union
half	largest	land features
South America	Brazilian	Amazon River Basin
3,900	coastal plain	climate

1. The largest country of South America is ____________________.
2. Brazil is nearly half as large as the entire continent of __________ __________.
3. Countries larger than Brazil are the __________ __________, __________, __________, and the __________ __________.
4. More than __________ of all the people of South America live in Brazil.
5. Where people live and work depends a great deal on __________ __________ and __________.
6. The largest region of Brazil is called the __________ __________ __________.
7. The Amazon River is __________ miles long and is the world's __________ river.
8. Three other land features (topography) of Brazil are a large __________ region, the __________ Highlands, and a narrow __________ __________.

Climates of Brazil. The equator and the Tropic of Capricorn both pass through Brazil. With that fact in mind, you can be sure that Brazil is a tropical country. The large lowland region near the equator is a huge forest land so large in area that it makes up nearly 40 percent of the country. The climate of the region is called *tropical rain forest.* Giant trees reach high for sunlight. Rainfall is heavy year round, often 80 inches or more in one year. Just about everything in the region is big except the number of people who live there.

A large part of Brazil has the *tropical savanna* climate. This climate is more favorable for people to live in than the rain forest. Many people feel that in years to come, the region will be very good for farming, cattle ranches, and other industries.

Along the southeastern part of Brazil, south of the Tropic of Capricorn, is the *humid subtropical* climate. The region is ideally suited for a great variety of farm crops and is not as tropical as other parts of Brazil. Although the coastal plain is very narrow in places, a number of large cities have grown in the region. The climate is hot in the summer and cooler in the winter. Rainfall is year round and sufficient enough for farming.

Directions: Fill in each blank with a term from the box. A word can used more than once.

Indians	40	Tropic of Capricorn
equator	rainfall	humid subtropical
trees	tropical savanna	tropical rain forest

1. Two very important lines of latitude that affect the climates of Brazil are the ____________________ and the ____________________________________ .
2. The large lowland region of Brazil has the ______________________________ climate.
3. The climate is hot all year round, has large ________ and plenty of __________ .
4. Nearly________ percent of Brazil has the tropical rain forest climate.
5. The number of people who live in the tropical rain forest climate is very small and mostly native __________________ .
6. Two other climates of Brazil are the______________________________ and the ____________________________________ .
7. Of the climates of Brazil, the ______________________________ appears to be the one that is most comfortable for people to live in.

Brazil—The Latex Boom. For a long time, Brazil has been a country famous for its Amazon River. In the mid 1800's, *latex*, a sap from a tree, was discovered in the rain forest region. Latex was used to make rubber tires and other products. Like the gold rush in California, the rush for latex made Brazil an important country. Unfortunately, the boom didn't last very long, and Brazil once again became less important. The latex period did, however, bring about some valuable information about the big rain forest region.

Brasilia—The New Capital. Officials of the Brazilian government decided to build a new capital city. Rio de Janeiro was the capital at that time and had a very large population. It was decided that an inland capital would be very good for Brazil. It would open up the large unsettled region and spread out the growing population. A *site* (place) 600 miles inland was chosen. Plans were made and in 1960, Brasilia became the new capital. New roads were built to the city which made traveling there faster and easier. Brasilia now has a population of over one million.

Directions: Fill in each blank with a term from the box. Some words may be used more than once.

boom	Brazil	600	Rio de Janeiro
latex	Amazon River	Brasilia	important

1. In earlier times, people knew about Brazil because of the__________ __________
2. In the mid to late 1800's, ________________ was discovered in the Brazilian rain forest.
3. As countries became more industrial, they needed more and more __________, a sap from a tree.
4. The rush to Amazonia, the rain forest, to get latex caused a __________.
5. The rush, or __________________, didn't last very long, and Brazil became less ______________________.
6. The new capital of Brazil is called ______________________.
7. The city is located__________ miles away from the sea coast.
8. Building the city inland would encourage (help) the growing population of __________________ to spread out.
9. The old capital of Brazil was______________________.

The Westward Movement. Ninety percent of the people of Brazil live within 200 miles of the coastal regions. The *vast* (large) interior has only a few people, most of whom are native Indians. The future of Brazil lies in greater use of those unsettled parts of the country. In the interior, the only major city other than Brasilia is Manaus. Manaus, which lies along the banks of the Amazon River, was developed during the latex boom. It now has a population of nearly 300,000. A city of that size in the heart of the rain forest is an important trade center for products of the rain forest. In recent years, an interest in promoting tourism has brought about modernization in an effort to make Manaus more attractive to visitors. In less than 25 years, with the government's help, more settlers are moving into the big empty lands of Brazil.

Brazil—Today. The needs of Brazil would make a long list. Improvements in transportation through better roads and railroads is one of these needs. More growth in manufactured products would be another. Increasing farm production is also important, as well as improving education and health care for the people.

A shortage in jobs, low pay for workers, over-crowded cities, and a fast growing population are all serious problems. To bring about improvements in all these areas and progress even further takes time and large sums of money, most of which are being loaned by other countries. Much progress has already been made, however, and Brazil has become a major industrial country in South America.

Directions: Complete the exercise.

1. In a brief paragraph, summarize the importance of the westward movement in Brazil.

____
____
____
____
____
____
____
____

2. Make a list of at least five problems that Brazil faces today.

   a. ____
   b. ____
   c. ____
   d. ____
   e. ____

3. Extra credit: Brazil has borrowed a great deal of money from other countries. Why is that a major problem?

____
____
____
____
____
____
____

ARGENTINA

The Geography of Argentina. Argentina is the second largest country in South America. In size, it is 1,065,200 square miles. The area would compare to that part of thc United States from the Atlantic Ocean to the Mississippi River. The population is about 31 million, the third largest in Latin America. Only Brazil and Mexico have larger populations. Argentina is in the Southern part of South America and has a variety of land features very much like the United States.

Land Features of Argentina. Argentina has five important regions. The Western part includes the high Andes Mountains. As you recall, the Andes run all the way to the Southern tip of South America. Mt. Aconcagua is a part of that region and is the highest mountain, not only in South America but the entire Western Hemisphere. This mountainous section has scattered valleys and plateaus. Its population is not very large.

A dry region lies east of the mountains. These drier sections of the region use irrigation to supply the necessary water for cattle, sugar cane, and grapes. Old established cities such as Cordoba, Mendoza, and Tucuman are located there.

North and east of the dry region is a wetter warmer section that produces a variety of crops. Corn, tea, flax (used for linen) cattle, and sheep are products of that area.

The heartland of Argentina is a region called the *Pampa*. It is located in the East Central part of Argentina. Lowlands extend nearly 400 miles inland. The farming season is year round with plenty of precipitation. It is Argentina's most productive and most populated region. Cattle, corn, and wheat are produced in very large quantities. The capital city of Buenos Aires is located along its coast.

Far to the south is a region called *Patagonia*. This large plateau region is dry and windy. Summers are hot, and winters are very cold. Where water can be found for irrigation, farmers can raise cattle, grow fruit, alfalfa (feed crop for animals), grapes and vegetables. In other parts, sheep are raised. Recent discoveries of oil and minerals could soon make Patagonia a more important region.

The Climates of Argentina. Argentina has climates very much like our country has. The highlands, as you know, are very much the same everywhere on the earth. The higher the *altitude* (height), the colder it gets. The importance of the Western Mountains of Argentina (the Andes) is the effect they have on precipitation. The *leeward* side (protected from the wind) will be drier.

The Northwestern region is a part of the savanna climate, a climate which has a season of rain and a season of dryness. Unless water can be provided by irrigation, farming is very limited. The Southern region of Patagonia is very dry and suitable only for grazing of sheep. Recent discoveries of oil and natural gas will make this area much more important than in the past.

The outstanding region of Argentina is the Pampa. The humid subtropical climate is very similar to the Southern part of the United States. Summers are hot, and winters are mild. Rainfall is year round and perfect for farming. The broad flatlands extend nearly 400 miles inland and is Argentina's most productive region.

Directions: Fill in each blank with a term from the box.

mild	Pampa	1,065,200
400	South America	Mexico
second	humid subtropical	rainfall
Patagonia	Brazil	31
United States	mountains	hot

Argentina is the _______ largest country in _____________ _____________. The area is________________ square miles with a population of about________ million. Only ______________________ and__________________ have a larger population in Latin America.

The major land features of Argentina are the Western _________________, a large plain called the ________________________, and a windswept region far to the south called ___________________________.

The climates of Argentina are very much like those of the _______________ __________________. The most productive region has a __________________ ___________________ climate. Summers are __________; winters are__________, and_______________________ is year round. Not only is the climate good for farming, but the region is very large and extends nearly_________ miles inland.

Industries of Argentina. Much of Argentina's business is carried on in a region called the Pampa. The capital city of Argentina, Buenos Aires, is located along the coast of the Pampa. It is a large city of nearly 3,000,000 people and a *surrounding* (metro) population of nearly 5 million more people, a total of over 8 million people.

Four-fifths of all the agricultural products are produced in the Pampa. Cattle, corn, and wheat are important agricultural products. Sheep, wool, dairy products, fruit, and vegetables are also important products. As in our southland, many farm crops are produced in great quantity.

In Argentina, auto assembly, flour mills, meat processing plants, textiles, chemical plants and machinery are all industries of importance. Recent discoveries of oil and natural gas, plus other minerals, will also help Argentina improve its economy.

Although Buenos Aires is Argentina's most important city, other large cities are, Tucuman, Cordoba, Bahia Blanca, and Mendoza. Study the map below.

Argentina

Argentina—Today. Argentina has rich farm lands, valuable minerals and skilled workers. There is little doubt that Argentina could become a leading country in trade and industries. A major problem has been political unrest in the country. Although the situation has now improved, there is still some question about the future.

Directions: Complete the exercise.

1. The heart land of Argentina is called the ____________________ .

2. The capital city of Argentina is ____________ ____________ .

3. Make a list of five important agricultural products of the Pampas region.

   a. ____________________ d. ____________________

   b. ____________________ e. ____________________

   c. ____________________

4. List five important industries of Argentina.

   a. ____________________ d. ____________________

   b. ____________________ e. ____________________

   c. ____________________

5. What are three advantages that Argentina has that could make the country a leader in trade and industries?

   a. ____

   b. ____

   c. ____

6. Extra credit: When a country has political problems, why does that affect the progress of that country? Write your answer on a separate sheet of paper.

Chapter 5
Asia

INTRODUCTION

Asia is the largest of the continents and has nearly two-thirds of the total world's population. The 17,250,000 square miles of land stretches one-third of the way around the earth. Very often Asia and Europe are combined as one continent and called *Eurasia*. However, in this regional study, Asia will be treated as a separate geographical region.

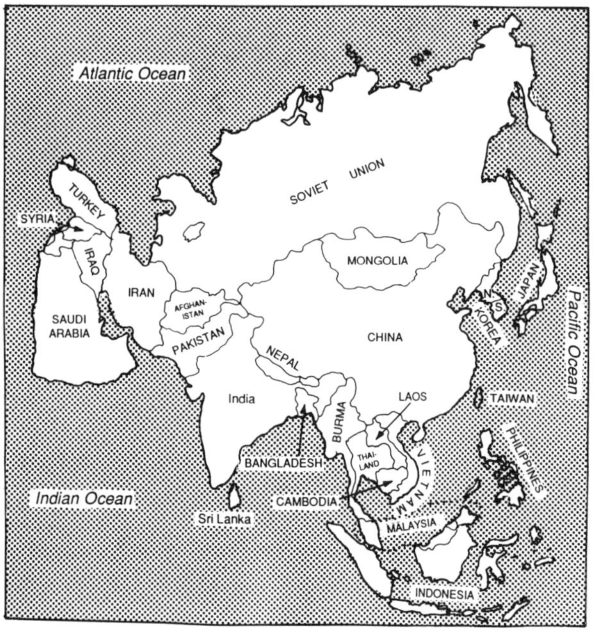

Asia

Land Features of Asia. The great size of Asia covers a range of latitude from near the equator (0 degrees) to well beyond the Arctic Circle ($66\,^1/_2$ degrees N.). Mountains, plateaus, valleys, lowlands, and coastal plains are all part of the land mass of Asia.

The highest mountain range on the earth is in this continent. The Himalayas of southern Asia are called the "roof of the world." Mt. Everest is over 29,000 ft. above sea level. That is nearly six miles high. Many other peaks of that mountain range are over 25,000 feet above sea level. The Himalayas are a very important land feature in Southeast Asia. Along with other important land features, the mountains greatly affect the climates, as well as where people live and the work they do.

Climates of Asia. As with the land features, almost every climate on the earth can be found in Asia. A land mass that covers such a great area has a great variety of climates.

The Southern part of Asia lies within the tropics. Its climates include the tropical rain forest, tropical savanna, and the tropical drylands. The middle latitudes have the *temperate* (moderate) climates. The humid subtropical, humid continental, and middle latitude drylands are found there. The taiga and tundra climates are in the Northern latitudes. Also very much a part of Asia is the varied mountain climate.

Directions: Fill in each blank with a term from the box.

Arctic Circle	largest	middle latitude drylands
29,000	taiga	tropical rain forest
Mt. Everest	Eurasia	Himalayas
population	humid subtropical	humid continental
tropical drylands	tropical savanna	climates
tundra	equator	

1. Asia is the ______________________ of all the continents in both area (size) and ______________________________________ .
2. Sometimes Europe and Asia are combined (joined) and called ________________ .
3. The land area (mass) stretches from near the _______________ (0°) to beyond the _____________________________________ (66 $^1/_2$ degrees North).
4. Because Asia is such a huge (large) land mass, many different _________________ are present.
5. The highest mountain range in the world is in Asia. It is called the ____________ .
6. ______________________ ______________________ is the highest peak and is over _________________ feet above sea level.
7. a. Asia also has a wide range of climates. The tropical climates are:

   ________________________________ ___________________________________

   ____

   b. Middle latitude climates are:

   ________________________________ ___________________________________

   ____

   c. High latitude climates are:

   ________________________________ ___________________________________

Japan

The Past. The early history of Japan is unusual in many ways. As an island country, very little was known about Japan. Strong *rulers* (emperors) would not allow anyone to come into the country. They did not want any changes.

The United States, however, had two problems with Japan. Ship-wrecked American sailors were not permitted to leave the country. President Millard Fillmore also wanted the Japanese to open their ports to trade. In 1853, President Fillmore sent Commodore Matthew C. Perry to visit Japan. Perry arrived in Tokyo Bay with a full squadron of armed ships. He presented the President's message to a high-ranking official. In 1854, he returned to Tokyo Bay, and the Japanese signed a treaty. Modern day Japan began at that time. The Japan of old was rapidly changed to a very Western-type of country.

Directions: In paragraph form, summarize the early history of Japan. In your paragraph, include each of the terms listed below. Underline each term. Write the paragraph on a separate sheet of paper.

emperors	ship-wrecked sailors	treaty
island country	1854	Tokyo Bay
President Fillmore	Commodore Perry	trade
1853		

Area and Population. Japan is a small island country made up of four major islands and many smaller islands. The area (size) of Japan is 145,856 square miles. The population is over 120 million. The area of the state of California is slightly larger and has about 27 million people. As you can see, Japan has a very large population for its size.

Land Features of Japan. The country of Japan lies off the eastern coast of Asia. The islands are volcanic mountains. There are some small lowland plains along the southern and eastern coasts of the largest island, Honshu (*HON shu*). In the northern island of Hokkaido (*HO ki do*) there is a lowland plain in the southwestern part of the island. Most of Japan, however, is very mountainous. Mount Fuji (*FU jee*) is the highest peak at 12,388 feet above sea level. Although some of the volcanic mountains are still active, Mount Fuji is not. Volcanic regions, such as Japan, have many earthquakes. The *tremors*, as they are called, are usually weak and harmless. Japan has over a thousand such tremors each year.

Climate of Japan. The four major islands of Japan extend from about 31 degrees N. latitude to 46 degrees N. latitude. Japan is in the same latitudes as the eastern part of the United States. Japan would extend from the state of Maine to the Southern part of the state of Georgia. The climates are also similar to that part of the United States.

Honshu, Shikoku (*Shi KO ku*), and Kyushu (*KYU shu*) are the three southern most islands of Japan. The climate of those islands, except in the mountainous parts, is humid subtropical. The summer season is hot and wet; the winter season is mild and rainy. The climate is very good for farming because of the long growing season and rainfall.

The northern island of Hokkaido has a much colder humid continental climate. Its summer season is hot and wet with a much shorter growing season. The winter season is cold and less wet. Farming can only take place during the warmer months of the year. All of the islands of Japan are mountainous. Therefore, as in all mountainous regions, the climate will vary depending upon the *elevation* (height). Other conditions will also affect the climate of those areas.

Directions: Fill in each blank with a term from the box. A term may be used more than once.

46	Asia	Maine	Hokkaido
island	Honshu	Kyushu	volcanic
31	Georgia	humid continental	12,388
Shikoku	humid subtropical	4	Mount Fuji

Japan is a small ________________ country in southeastern ________________. There are ________ major islands, the largest of which is ________________. Most of the land of Japan is made up of ____________________ mountains. The highest peak in Japan is ____________________, at ______________ feet above sea level.

Japan extends from about ________ degrees north latitude to ________ degrees north latitude. The latitude of Japan when compared to our country, would be from ______________ to southern ________________. The three most southern islands of Japan are ____________, ________________, and ________________. The most northern island is called ________________. Southern Japan has the ________________________________ ____ climate. Northern Japan has the ____ ____________________________ climate.

Changes from Old to New. An important date in the history of Japan was 1854. Once Japan opened its doors to the rest of the world, changes took place quickly. The agricultural way of life began to change. Improvements were made in roads, railroads, education, communication and industries. Japanese students traveled to Western countries to learn new ideas. Many countries showed Japan how to build factories and improve their farming methods. In less than two generations (50 years), Japan had become the most advanced country in Asia.

Japan At War. At one time or another from 1900 to 1945, Japan was involved in wars with Russia, China, and the United States. In 1945, Japan was defeated and forced to surrender to the United States. The empire that Japan had built during that time no longer existed. After those war years, Japan quickly recovered and once again became the leading industrial country of Asia.

Farming In Japan. Japan is a mountainous country with very little good land for farming. Some small lowland regions are found on the islands of Hokkaido and Honshu. To get more land for farming, hilly regions are *terraced.* Terracing is done by notching out level pieces of land. Step-like sections of level land extend up the hillside. No land is wasted.

The Japanese are excellent farmers. The yield per acre is very high because the newest methods are used. The large population of Japan requires a great deal of food. Fishing provides a source of food supply. The coastal waters are rich in a great variety of fish. Japan, however, must still *import* (buy) food from other countries.

Industrial Japan. The change from an unknown country to a great industrial country began in 1854. Once Japan opened its doors to the outside world, rapid changes took place. Japan sent bright students to all parts of the world to learn as much as they could. Other countries were invited to help Japan build railroads, roads, hospitals, schools, and factories. Japan copied those ideas and soon became the most industrial nation in Asia.

Trade increased between Japan and many of the more industrial countries. Japan grew larger. At first, Japanese products were said to be "not as good, but twice as cheap." Since 1945, however, that old saying no longer applies. The leading industries of Japan today are steel, automobiles, shipbuilding, *textiles* (cloth), electronic equipment, cameras, silk, and electrical products.

Japan is the third largest industrial nation in the world. Only the United States and the Soviet Union are more industrial. Along with Japan's great success, also comes problems. Crowded cities, pollution, and the need for trade to survive are serious matters. Hard working Japanese, however, will no doubt find the answer to these problems.

Directions: Complete the exercise. Answer each question in a complete sentence.

1. Why was 1854 an important date in Japan's history? ____________________

   ____

2. How long did it take Japan to change from an agricultural (farming) country to an industrial country? ____________________

   ____

3. Why was 1945 an important date in the history of Japan? ____________________

   ____

4. Why is farming so difficult (hard) in Japan? ____________________

   ____

5. What is meant by *terracing*? ____________________

   ____

6. Why is fishing important to Japan? ____________________

   ____

7. Japan copied ideas from other countries. How did that help Japan to make progress? ____________________

   ____

8. Make a list of seven important industries of Japan.

   a. ____________________ e. ____________________

   b. ____________________ f. ____________________

   c. ____________________ g. ____________________

   d. ____________________

9. How does Japan rank as an industrial country? ____________________

   ____

10. List three serious problems of present day Japan.

    a. ____

    b. ____

    c. ____

Chapter 6
Europe

Introduction

Compared to the other continents of the earth, Europe is rather small. Only Australia is smaller in size. Many people are surprised by the fact that Europe is just a little larger than the United States. The area of Europe is 4,063,000 square miles. The area of the United States is about 3,615,000 square miles. The small continent of Europe, however, has the second largest population of all the continents. Only Asia has more people. Europe has been the center of great explorers, ideas, and culture for hundreds of years. Brave European explorers discovered the lands we call North America and South America. The history of our own country dates back to early English colonies. Later, millions of other Europeans chose to come to our country. Europe of the past and of the present is a very important place to us and to the rest of the world.

Directions: Fill in each blank with a term from the box.

North America	South America	Australia
population	small	Europeans
4,063,000	English	3,615,000

1. Compared to other continents, Europe is rather ____________ .
2. Only the continent of ____________________ is smaller in area than Europe.
3. The land area of Europe is about ______________ square miles.
4. The land area of the United States is about ______________ square miles.
5. Europe has the second largest ____________________ of all the continents.
6. European explorers discovered the continents of ____________ ____________ and ____________ ____________ .
7. The history of our country dates back to early ______________ colonies.
8. Millions of other ____________________ came to our country.

Europe

Peninsulas of Europe. Europe is made up of several *peninsulas.* A peninsula is land which is bordered on three sides by water. See map above. The Scandinavian Peninsula, the Iberian Peninsula, the Italian (Apennine) Peninsula, the Jutland Peninsula, and the Balkan Peninsula are all shown on the map. With so many peninsulas, no part of Europe is very far away from a major body of water. It is easy to see why European countries were so interested in exploration and trade.

Land Features of Europe. Much of the continent of Europe is made up of mountainous land. Northern Europe and southern Europe are regions of highlands. The Alps of southern Europe are probably the most famous mountains in the world. They are not the highest, but they are best known. The only major lowland region is called the *Great European Plain.* The plain is very important to northern France, Denmark, and northern Germany. Part of the plain extends into southern Sweden.

Mountains, plateaus, and smaller lowland regions, however, make up most of Europe. The land has been used wisely. The *favorable* (good) climates also help to make Europe a very important continent.

Directions: Write the name of each peninsula. Then write the country or countries that the peninsula contains. Use the map on page 47.

	Peninsula	**Country or Countries**
a.	____________________	____
b.	____________________	____
c.	____________________	____
d.	____________________	____
e.	____________________	____

THE CLIMATES OF EUROPE

The climates of Europe are generally warmer in the winter and cooler in the summer because of the large bodies of water nearby. The warm Gulf Stream flows along Western Europe. It also keeps Western Europe milder in the winter time.

The Marine West Coast Climate. The marine west coast climate is very important to Western Europe. The westerly wind blows over the warm Gulf Stream. Parts of the British Isles, France, northern Spain, western Norway, Denmark, Belgium, the Netherlands, Luxembourg, and West Germany have the marine west coast climate. The climate goes far inland because there are no mountains to block the winds. Summers are cool and rainy, winters are mild and rainy. Many people live in this part of Western Europe.

Taiga and Tundra Climates. Europe extends to 71 degrees N. latitude. The Arctic Circle is 66 $^1/_2$ degrees N. Regions that far north are very cold. Summers are short, and winters are very long. Not many people of Western Europe live in those climates.

The Mediterranean Europe. The narrow coastal regions of southern Europe have the Mediterranean climate. The summers are hot and dry, and the winters are very mild and rainy. Water is the big problem in the Mediterranean climate. The mountainous regions nearby have plenty of water. The key to success in those Mediterranean regions is being able to get the water from the mountains to the coastal plains. Large dams and reservoirs cost a great deal of money to build. Pipelines are also needed to carry the water to farms and cities. The Mediterranean climate is outstanding if water can be made available.

A. Directions: Describe each climate listed below.

1. Marine west coast climate ______________________________

____

____

2. Taiga and tundra climates ______________________________

____

____

3. Mediterranean climate ______________________________

____

____

FRANCE

The Crossroads of Western Europe. The area (size) of France is 221,207 square miles, slightly smaller than the state of Texas. The population is about 56,343,000. With the exception of the Soviet Union and Turkey, France is the largest Western European country in size. The location of France is excellent for trade. The southern part of France borders along the Mediterranean Sea. To the west and northwest are the Bay of Biscay, the Atlantic Ocean, and the English Channel. It is easy to see why France has been called the "crossroads of Western Europe."

B. Directions: Complete the exercise.

1. What is the area of France? ______________________________

   What is the population? ______________________________

2. How does France compare in size to other European countries?

____

3. Bodies (or body) of water that border France:

   a. to the south is ______________________________ .

   b. to the west and northwest are ______________________________ .

4. Why is France called the "crossroads of Western Europe"? ______________________________

____

____

Land Features of France. Along with the excellent location is the large lowland plain of western and northern France. More than half of France is a fertile plain crossed by many rivers. The rivers and man-made canals are very important. Products and other goods can be shipped to all parts of France. The land is much higher in the east and south. The eastern part is made up of lower mountains, but the mountains to the south are higher. The Pyrenees Mountains separate France from Spain. The Alps in the southeast separate France from Italy and Switzerland. Its location, large lowland plain, excellent rivers, and mountainous regions are very important to France and Western Europe.

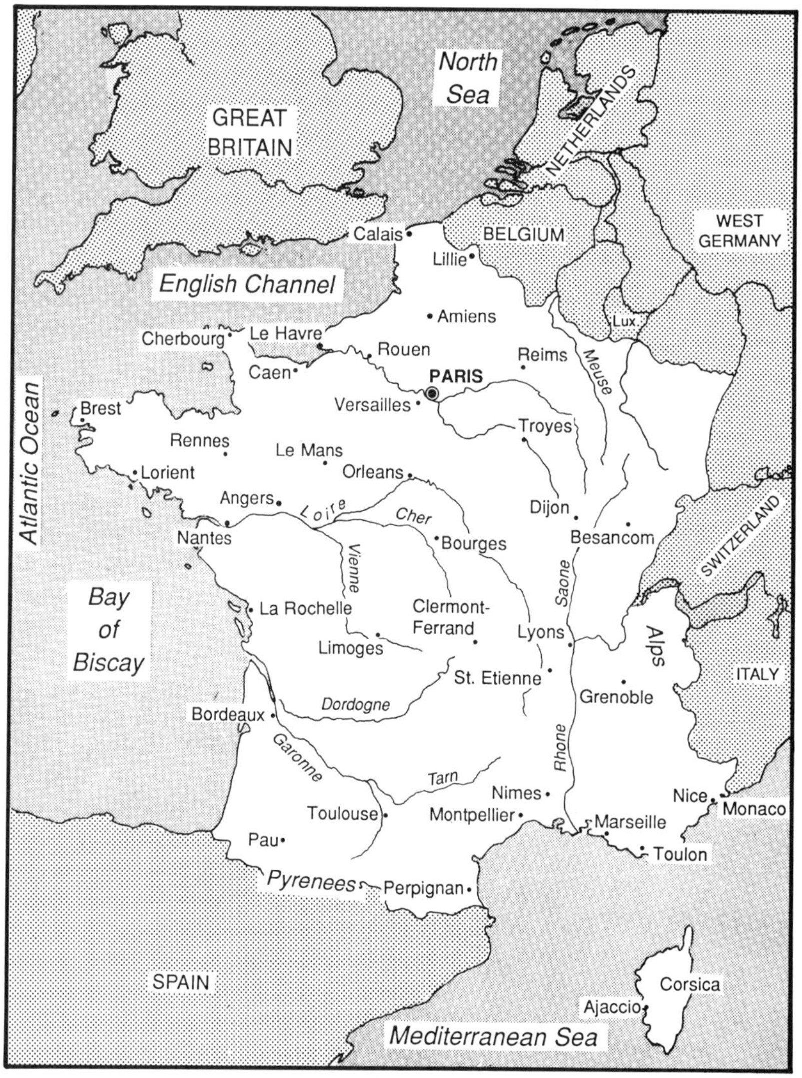

France

Climates of France. France is bordered by large bodies of water. The Atlantic Ocean and the Bay of Biscay lie to the west and the Mediterranean Sea to the south. The westerly winds over the Atlantic Ocean blow inland. In this marine west coast region, summer is warm and comfortable. High temperatures are unusual. Rainfall is frequent and good for farming. The winter season is cool with light rains. Very cold temperatures are also unusual because of the westerly winds from the large ocean. The Mediterranean climate is found along the narrow coastal plain of southern France. There, the summer season is hot and very dry. The winter season is mild and rainy. The high mountains nearby provide much-needed water for homes, farming, and industries.

Geographically, France has many advantages. The good location, comfortable climates, lowlands, excellent rivers, long coastline, and beautiful mountains help make France a modern country.

A. Directions: Write *True* or *False* for each statement.

__________ 1. France has an excellent location.

__________ 2. More than half of France is a fertile lowland plain.

__________ 3. Many rivers cross the lowlands of France.

__________ 4. A canal is a man-made waterway.

__________ 5. Rivers can be used to ship goods from one part of France to another.

B. Directions: Fill in the blank. Choose your answer from the words below.

Switzerland	plain	Alps	Pyrenees
location	rivers	Italy	France

The ________________ Mountains separate ________________ from Spain. The ________________ separate France from ________________ and ________________ . Three important advantages that France has are an excellent ________________ , a large lowland ________________, and excellent ________________ .

C. List three large bodies of water that border France.

1. ________________ 3. ________________

2. ________________

D. List the two major climates of France. Write one fact about each.

1. __________ ____________________

2. __________ ____________________

Farming. A very large part of France is used for farming. In the past, farms have been small, usually 30 - 40 acres. In more recent times, however, farms have become larger. With larger farms, more farm machinery is being used. The French are very proud of their land and are very good farmers. France is the only country of Western Europe that can produce nearly enough food for its people. Farm products are also sold to other European countries. A great variety of crops are produced, such as potatoes, sugar beets, barley, rye, oats, olives, oranges, and grapes. Livestock is also raised in large numbers.

Industries. France also has many other important industries. French wines, for instance, are world famous. Different types of grapes are grown which produce quality wines. Winemaking is one of France's important industries. For many years, France has also been known as a country that makes very special and expensive products. Some of these are perfumes, clothes, jewelry, pottery, and fine silks. Now, France has added automobiles, airplanes, railroad equipment, aluminum products, farm equipment, and other machinery to its growing list of products.

The most important city, and the capital of France, is Paris. It has over two million people. Le Havre (*luh - AH - vruh*), Bordeaux (*bawr - DOH*), Toulouse (*too - LOOZ*), Marseille (*mahr - SAY*), Lyons (*lee - AWN*), and Cherbourg (*SHARE - boorg*), are also important cities. On the island of Corsica, which belongs to France, is an important city called *Ajaccio*. It is the capital of Corsica and the birthplace of Napoleon Bonaparte, once Emperor of France.

Although, France has changed in many ways, the small villages, small farms, hard-working farmers, and quiet life are still very much a part of France today.

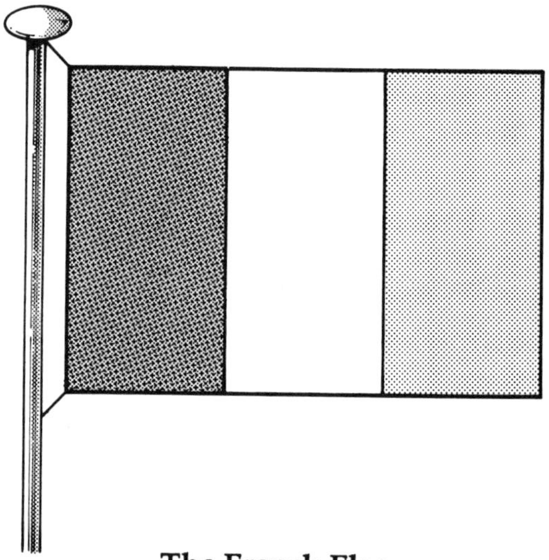

The French Flag

A. Directions: Underline the best answer.

1. A (large, small) part of France is used for farming.
2. Usually, farms in France were (30-40 acres, 100-200 acres).
3. Larger farms can use (more, less) machinery.
4. The farmers of France produce (many, few) different kinds of crops.
5. As an industry, farming is (important, not important) in France.
6. The capital city of France is (Paris, Ajaccio).

B. Directions: List six products that France produces.

1. ______________________ 4. ______________________
2. ______________________ 5. ______________________
3. ______________________ 6. ______________________

C. Directions: List six important cities of France.

1. ______________________ 4. ______________________
2. ______________________ 5. ______________________
3. ______________________ 6. ______________________

D. Directions: Fill in each blank.

The important city on the island of Corsica is ________________. It was the birthplace of ________________ ________________.

Chapter 7
Africa

INTRODUCTION

Africa is the second largest continent. Its area is about 11,707,000 square miles. With over 600 million people, it ranks third in population. Only Asia and Europe have larger populations than Africa. Africa's population is similar to that of southern Europe and Asia.

The coastline of Africa was explored by a Portuguese sailor, Vasco da Gama, in the late 15th century. Hundreds of years passed before European explorers went into the *interior* (inside) of Africa. With so little known about Africa, it was called the "Dark Continent."

Directions: Complete the exercise.

1. What is the area of Africa? ______________________________

   What is the population? ______________________________

2. How does Africa compare in area and population with other continents?

   ____

3. List the two continents that are close to Africa.

   a. ____________________ b. ____________________

4. Who was the Portuguese sailor that explored the coastline of Africa?

   ____

   When? ______________________________

5. Why was Africa called the "Dark Continent"? ______________________________

   ____

Africa

Land Features of Africa. Africa is made up mostly of a very large plateau which is lower in the north and higher in the south. Scattered mountains are found throughout the plateau. The only range of mountains is found in northwestern Africa. The Atlas Mountains are rather close to the coasts of the Atlantic Ocean and the Mediterranean Sea. The edges of the plateau drop sharply to the coastline. The coastal plain is very narrow, and the coastline has very few good harbors.

Crossing the plateau are several large rivers. The Nile, Congo, Niger, Zambezi, Limpopo, and Orange Rivers flow through various parts of Africa. The Nile is the longest river, over 4,100 miles in length. Giant waterfalls are on those rivers as they drop from one height of the plateau to another. The waterfalls make it hard for ships to travel very far on the rivers. The only lowland regions are found in western and northeastern Africa.

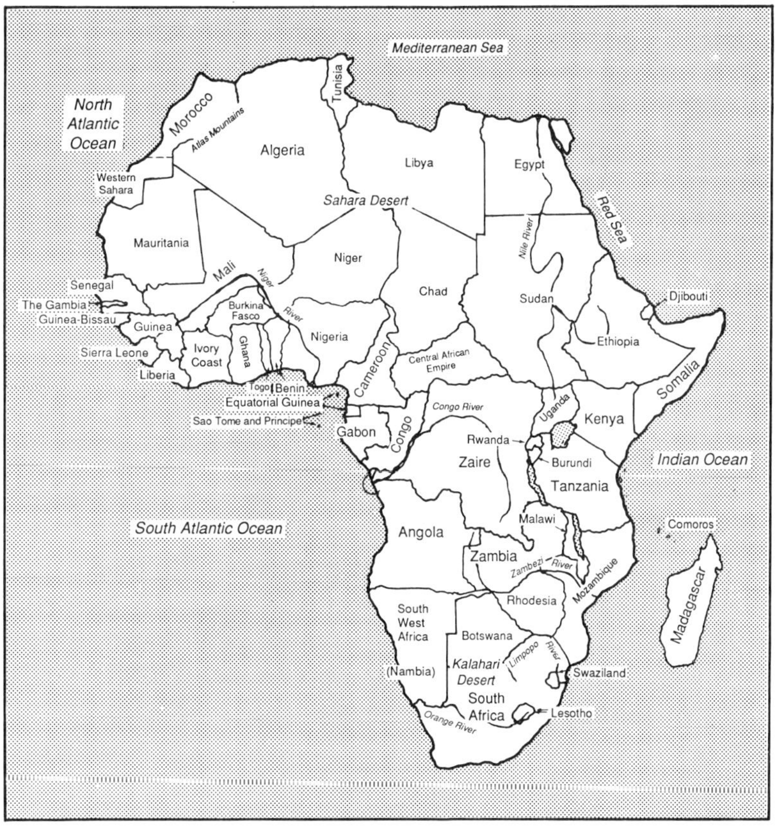

Africa

Directions: Fill in each blank with a term from the box.

longest	Atlantic	plateau
western	harbors	Nile
Atlas	Zambezi	northwestern
Niger	Orange	northeastern
4,100	waterfalls	narrow
ships	Mediterranean	Congo

Most of Africa is made up of a very large ______________ . The ______________ Mountains are found in ______________________________ Africa. The mountains are very close to the coasts of the ______________________________ Ocean and the ______________________________ Sea. The plateau drops sharply to the coast. The coastal plain is very ______________________ , and the coastline has very few good ______________________ . Important rivers that cross Africa are the ______________________ , the ______________________ , the ______________________ , the ______________________ , and the ______________________ . The Nile River is the ________________ river in the world. It is over __________ miles long. Along those great rivers are many ______________________ . The waterfalls make it hard for ________________ to travel on the rivers. The only lowland regions are found in ______________________________ and ______________________________ Africa.

Climates of Africa. The location of Africa is about 38 degrees N. latitude to 35 degrees S. latitude. The equator (0 degrees) divides Africa into two nearly equal parts. The Tropic of Cancer $23\,^1/_2$ degrees N., and the Tropic of Capricorn $23\,^1/_2$ degrees S., also pass through Africa. A large part of Africa lies either in the tropical zone or very near it. With the exception of mountains or very high plateau land, the climates of Africa are warm year round.

The tropical rain forest climate is along the equator. The Congo River flows through the region. The climate is hot and wet throughout the year. Bordering the rain forest is the tropical savanna climate. The *savanna* (grasslands) is a very large area. The climate has two seasons. Both seasons are hot, but one is dry, and the other is wet. Tropical desert regions are north and south of the grasslands. The largest and most famous of all deserts is the huge *Sahara.* It is nearly as large as the entire United States, three and a half million square miles in area. The *Kalahari* desert lies in the southern part of Africa. The deserts are hot most of the year and very dry. Nighttime temperatures can drop quickly and sometimes even go lower than 32 degrees F. The three tropical climates are not well suited for large populations.

The Mediterranean climate is found along the narrow coastal plains of northern and southwestern Africa. The Mediterranean climate has a hot, dry summer and a mild wet winter. Nearby mountains provide water for the coastal cities and irrigation for farming. It is a very attractive climate, very similar to southern California in our country. The climate attracts a large population if a large enough supply of water is available.

In the southeastern part of Africa is the humid subtropical climate. The climate is excellent for farming. The summer season is hot and wet, and the winter is mild and has frequent rains. The climate is well suited for a large population.

Directions: Complete the exercise.

1. List the three important lines of latitude that cross Africa. Also, write the degree marking for each line.

   a. ______________________ c. ______________________

   b. ______________________

2. Complete the chart. One has been done for you.

Climate	Location	Seasons
a. *tropical rain forest*	*central*	*hot, wet*
b.		
c.		
d.		
e.		

3. List the two tropical deserts of Africa.

   a. ______________________ b. ______________________

4. What does the word *savanna* mean?

   ____

5. What is the big problem with the Mediterranean climate?

   ____

6. Which climate of Africa is best suited for farming? ______________________

____

Ghana

Ghana. The continent of Africa is often divided into two parts. One part is north of the large Sahara Desert and borders along the Atlantic Ocean and the Mediterranean Sea. That part of Africa is made up largely of *Caucasian* (white) people. South of the Sahara is a region called *sub-Sahara* Africa. Sub-Sahara means below or south of the Sahara Desert. That region is made up largely of black Africans. One of the countries of sub-Sahara Africa is Ghana.

Ghana

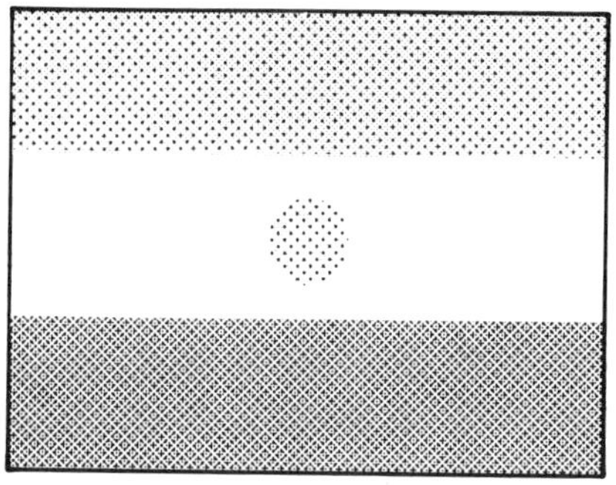

Niger

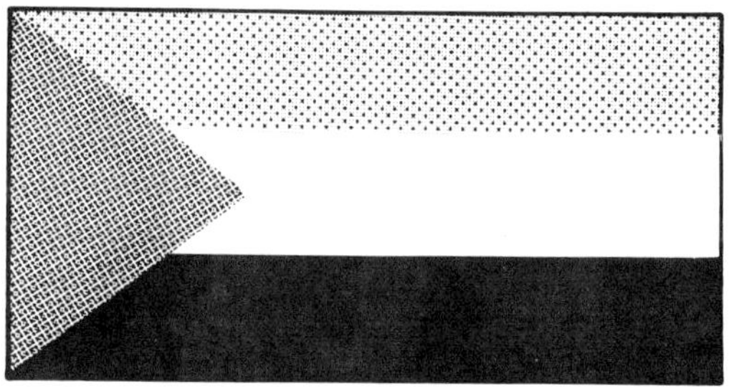

Sudan

Sao Tome and Principe

Land Features of Ghana. Ghana is a country located along the western coast of Africa. The area (size) is over 92,000 square miles, about the size of the state of Oregon. The population is about 13 million. Ghana is bordered to the west by the Ivory Coast, to the north by Burkina Faso (formerly Upper Volta 1984), and Togo to the east.

A large part of Ghana is made up of land 2,000 - 3,000 feet above sea level. The lowlands of Ghana are located along the coast near the capital city of Accra.

The coastline of Ghana is very straight. A straight coastline is not good for harbors because the water is often not deep enough to take large ships.

Climates of Ghana. Ghana is close to the equator (0 degrees). Accra, the capital city, is about 6 degrees from the equator. The two climates of Ghana are both tropical. A small part in the southwest is tropical rain forest. The climate only has one season, which is hot and wet. The other climate is tropical savanna. Savanna (grasslands) has two separate seasons. Half of the year it is hot and very wet. The other half of the year is warm and nearly dry.

Directions: Complete the exercise.

1. What are the two parts of Africa? ______________________

 ____

2. What separates the two parts of Africa? ______________________

3. Where is the country of Ghana located? ______________________

 ____

4. What is the size of Ghana? ______________________

 What is the population? ______________________

5. List the country that borders Ghana to the:

 a. west ______________ c. north ______________

 b. east ______________

6. What is the capital city of Ghana? ______________________

7. What are the two climates of Ghana? Write one fact about each of the climates.

 a. ______________________

 ____

 b. ______________________

 ____

Ghana—Today. For over a hundred years, Ghana was known as the *Gold Coast*. The name was used because gold was discovered in the region. At one time, the British government *ruled* (controlled) the colony. In 1957, the Gold Coast was the first of the British colonies in Africa to become independent. The name Gold Coast was changed to Ghana.

Ghana's major industry is farming. The most important crop is *cacao*. The cacao bean is used to make chocolate. Cotton, coffee, latex (rubber), palm oil, and special trees are also important products. A large dam was built on the Volta River for water power and to improve farming. The dam stores water during the wet season. During the dry season, the water is used to *irrigate* (wet) the land.

Ghana also has some important minerals. The two important minerals are manganese and bauxite. Manganese is used with other metals to make steel. Bauxite is used to make aluminum. Gold and diamonds are also *exported* (sold) to other countries.

Since 1957, the new government of Ghana has been faced with many problems. At various times over the past years, the government has also been taken over by strong military *leaders* (dictators). Changes in the government occurred over the years. These changes have led to more problems. Ghana could be the leading country of West Africa. A *stable* (no changes) government, therefore, is very important to the future success of Ghana.

Directions: Use the information above to complete the exercise.

Ghana was once called the __________ ____________. For over a hundred years, Ghana was a colony controlled (ruled) by the ___________. In the year ______, Ghana became an independent country.

The most important industry of Ghana is ____________. The important crop is ____________ which is used to make ____________________. Some other important crops are ____________________, ______________, __________, and ______________ ______________.

A large dam was built on the ______________ River. During the dry season, water from the lake behind the dam is used to ______________________ the land.

Ghana also has some very important minerals. ____________________ is used to make steel, and __________________ is used to make aluminum. ____________ and _________________ are also sold to other countries.

Ghana could be the leading country of Western Africa. The big problem has been the frequent changes in the ______________________.

Directions: Choose terms from the box to match each heading below. One has been done for you. All of the lines will not be used.

Atlas	Nile	Ivory Coast
Atlantic	Zambezi	Burkina Faso
Mediterranean	Orange	Togo
Congo	Sahara	Accra
Volta	Ghana	Kalahari

Country

Togo

City

Desert

River

Water Body

Mountain

Directions: Use the information that you have learned to write a summary paragraph about Ghana. Be sure to use these key terms in the paragraph: *Accra, Volta River, cacao, 1957, independent, colony, British,* and *Gold Coast.* The terms can be used in any order.

Chapter 8
The Middle East

Introduction

The Middle East, or Near East as it was called, is a part of the *land mass* (continent) of Asia. It is located at the far eastern end of the Mediterranean Sea. Europe and Africa border the Middle East to the northwest and southwest. The countries of the Middle East include Turkey, Syria, Lebanon, Israel, Jordan, Cyprus, Iraq, Iran, Egypt, Saudi Arabia, Yemen Arab Republic, People's Democratic Republic of Yemen, Oman, United Arab Emirates, Qatar, Bahrain, and Kuwait. In earlier times, the location of the Middle East was the crossroads of land and water travel. Today, the location continues to be of great importance. The Suez Canal connects the Red Sea with the Mediterranean Sea. The Gulf of Suez, as shown on the diagram, is an arm of the Red Sea. Ships of many countries use the canal as a short cut to Europe, Asia, and North America.

In addition, the region has a very large supply of *petroleum* (oil). Many countries, including the United States, buy millions of barrels of oil from the Middle East. Oil is used to make gasoline. Oil is also used to run machinery, heat homes and factories, and to generate electricity. The United States depends very heavily on oil from the Middle East.

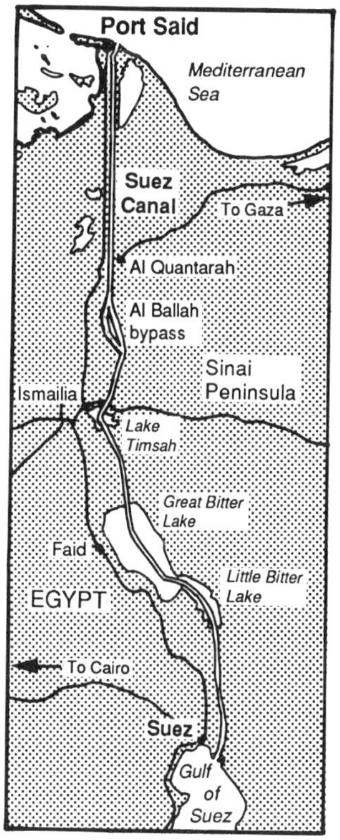

Suez Canal

There are many political problems in the Middle East. Countries within the region have been engaged in constant battle. The area has been a site of much unrest.

Directions: Complete the exercise.

1. List the countries of the Middle East. Number each country. There are 17 in all.

2. Where is the Middle East located?

3. Why is the location of the Middle East important?

Land Features of the Middle East. The Middle East is a region of very little lowlands except along the coasts and river valley. A large part of the region is made up of plateaus and mountains. In addition to the rugged land, a large part of the region is very dry.

Climates of the Middle East. A large part of the Middle East has the tropical desert climate. The desert regions of Egypt and the Arabian Peninsula are *arid* (have very little rain). These regions have less than 10 inches of precipitation a year. There are some regions that are *semiarid* (have light rainfall). The semiarid regions have 10-20 inches of precipitation a year. The Mediterranean climate is found along the coasts of the Mediterranean Sea. The climate has a hot, dry summer and a mild, rainy winter. The northern and eastern parts of the region have the *middle latitude steppe* climate. The steppe climate is semiarid. Winters are much colder than the winters of the Mediterranean climate. In the mountainous regions, the temperatures are generally colder. Precipitation varies from one part to another.

Directions: Complete the exercise.

1. List three land features of the Middle East.

   a. ______________________________

   b. ______________________________

   c. ______________________________

2. List five climates found in the Middle East.

   a. ______________________________ d. ______________________________

   b. ______________________________ e. ______________________________

   c. ______________________________

3. Define each term.

   a. arid ____

   b. semiarid ____

Saudi Arabia

Saudi Arabia is the largest country in the Middle East and is located on the Arabian peninsula. The area is about 865,000 square miles, about one-fourth the size of the United States. The population is over 11 million. Saudi Arabia makes up most of the Arabian peninsula, which is the largest peninsula in the world. There are several smaller countries on the peninsula. To the south of Saudi Arabia are Yemen (North Yemen), the People's Democratic Republic of Yemen (South Yemen), and Oman. To the north are Kuwait, Jordan, and Iraq. To the east are Qatar, Bahrain and the United Arab Emirates. To the west and northwest are Egypt and the Red Sea. In 1932, Saudi Arabia became an independent country. The government is a *monarchy* (ruled by a king). Most of the people of Saudi Arabia follow the Islamic religion. They are called *Muslims*.

A. Directions: Fill in each blank with a term from the box.

Muslims	Middle East	1932	largest
11	Islamic	865,000	Saudi Arabia

The Arabian Peninsula is the ________________ peninsula in the world and is located in the ________________ ________________ . The country of ________________ ________________ makes up a large part of the Arabian peninsula. It is ____________ square miles in size and has a population of about __________ million. In the year of ________ , Saudi Arabia became an independent country. Most of the people of Saudi Arabia follow the ______________ religion. They are called ________________ .

B. Directions: List the countries that are neighbors of Saudi Arabia under the correct heading.

South	North
____________________	____________________
____________________	____________________
____________________	____________________

East	West
____________________	____________________
____________________	____________________
____________________	____________________

Land Features of Saudi Arabia. From north to south along the Red Sea is a very narrow coastal plain. East of the coastal plain is a mountain range. The mountains run north to south for 1,000 miles. Beyond the mountains is a large inland plateau. The plateau covers most of Saudi Arabia. The plateau slopes downward to the eastern border along the Persian Gulf. The major cities of Jidda, Mecca, and Medina are located in the western part of Saudi Arabia. The capital city of Riyadh (*ree YAHD*) is located far inland, about 250 miles from the Persian Gulf.

Climates of Saudi Arabia. Saudi Arabia is very dry. The country is so dry that rivers dry up and flow only during a short rainy period. In southwestern Saudi Arabia, there is a little more precipitation. Even that region is still considered to be dry. The tropical desert is hot and dry for many months of the year. As in all tropical deserts, nighttime temperatures drop quickly. The cooler months are still warm. There is enough rainfall during those cooler months for grass to grow. The mountainous regions are much cooler and wetter.

Directions: Complete the exercise.

1. List the three major land features of Saudi Arabia.

   a. ______________________ c. ______________________

   b. ______________________

2. Of the three land features, which one covers most of Saudi Arabia? ____________

3. List four important cities of Saudi Arabia.

   a. ______________________ c. ______________________

   b. ______________________ d. ______________________

4. What is the capital city of Saudi Arabia? ______________________

5. List four important facts about the tropical desert climate.

   a. ______________________ c. ______________________

   b. ______________________ d. ______________________

Special Question:

6. Why are mountainous regions cooler and wetter in Saudi Arabia?

____

____

____

Saudi Arabia—Farming. Farming in Saudi Arabia is not easy to do. There is not much fertile land and there is very little rainfall. Some parts of the country have an *oasis*. An oasis is a place where underground water comes to the surface or near the surface of the land. Wells can be dug, and the water can be pumped to the fields. With irrigation, some crops can be grown. In the southwestern part of the country, coffee, wheat, and barley are grown. Dates are also grown in the eastern part of the country near the Persian Gulf. Farming however, is very limited. Sheep, goats, horses, and camels are raised in various parts of the country where there is an oasis. Sheep, goats, and camels provide many of the daily needs of the herders. The camel also is a strong animal and can be used to carry heavy loads. Camels also can go for days without water. With more dams and reservoirs being built, more water can be used to irrigate larger areas of the country.

A Camel

Directions: Complete the exercise.

dry, oasis, wells, irrigation, dams, reservoirs

1. In paragraph form, use each of the above words to summarize farming in Saudi Arabia. You may use the words in any order. Use your own paper.
2. List three important crops grown in Saudi Arabia.

   a. ______________________ c. ______________________

   b. ______________________
3. Why are sheep, goats, and camels raised in Saudi Arabia? ______________________

   ____
4. What are two facts about camels?

   a. ____

   b. ____

Saudi Arabia—Oil. The discovery of *petroleum* (oil) along the Persian Gulf has made Saudi Arabia a very important country. The oil fields in Saudi Arabia are able to produce millions of barrels of oil every day. Petroleum is used to make gasoline and other needed oil products. Many industries use oil as a fuel to run the factories. Oil is also used to generate electrical power, as well as to heat buildings and homes. Many countries in the world need a very large daily supply of oil. Our country uses millions of barrels of oil every day. Along with other oil-producing countries of the Middle East, Saudi Arabia has raised the price of oil. How much oil costs is important to our country and other industrial countries as well. Without oil, many of the industrial countries would be faced with a serious problem.

How Saudi Arabia spends the money made from selling oil is very important to the future of that country. Roads, railroads, schools, hospitals and new industries are needed. New industries would provide many new jobs. These new industries would be even more important if the supply or demand for oil runs out. Although some improvements have already been made, much more needs to be done.

Directions: Complete the exercise.

1. Why is Saudi Arabia an important country? ______________________________

____

2. List three reasons why petroleum is an important mineral to many countries.

   a. ____

   b. ____

   c. ____

Special Question:

3. Why is the cost of oil important to our country and other countries as well?

____

____

4. Why would new industries be of great importance to Saudi Arabia in the years to come? ____

____

____

Directions: Write the capital of each country given below. Use the map to complete the exercise. All of the countries are not listed. The map also shows the different climates of the Middle East.

The Middle East

Country	Capital
1. Turkey	____________________
2. Syria	____________________
3. Lebanon	____________________
4. Israel	____________________
5. Jordan	____________________
6. Cyprus	____________________
7. Iraq	____________________
8. Iran	____________________
9. Egypt	____________________
10. Saudi Arabia	____________________

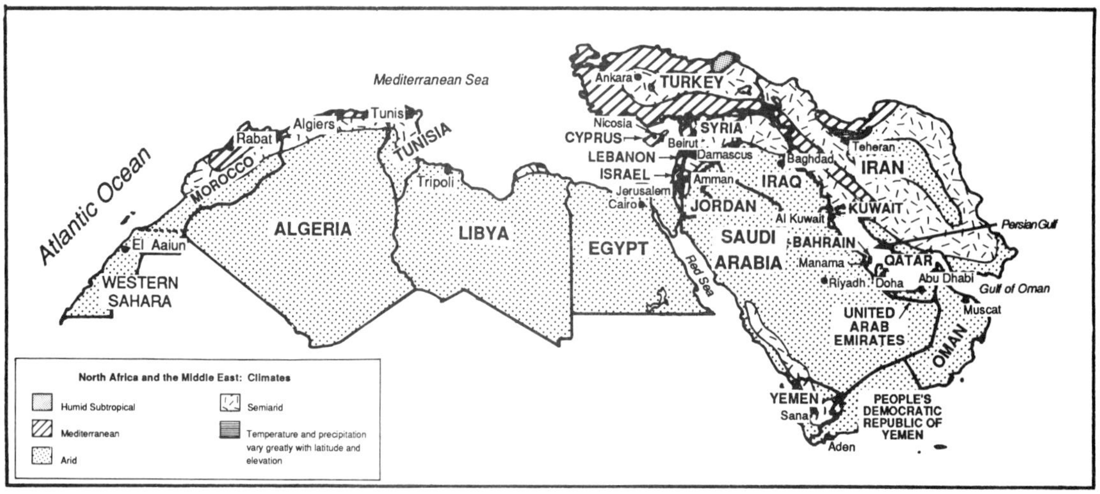

The Middle East

Chapter 9
North America

Introduction

North America is a very large continent, over 9,400,000 square miles in area. Asia and Africa are the only continents larger than North America. North America extends from the Arctic Ocean to the Gulf of Mexico and the Atlantic Ocean to the Pacific Ocean. Two of the largest countries on the earth are located in North America. They are Canada and the United States. Mexico and the countries of Central America are also included in North America. Geographically, however, Mexico and Central America are usually included as part of Latin America.

Land Features of North America. Canada and the United States make up most of North America. Towering mountains, large plateaus, great lowland plains, river valleys, wide, as well as narrow coastal plains and low mountains are all land features of the two large countries. The land varies in height from Mt. McKinley in Alaska, 20,320 feet above sea level, to Death Valley in California, 282 feet below sea level.

New York Harbor: The Statue of Liberty

Climates of North America. The large continent of North America extends from about 82° north latitude to about 9° north latitude. The spread of latitude from beyond the Arctic Circle, (66 $^1/_2$° N.) to almost the equator (0°) produces many climates. All of the climates on the earth are found in North America. The *frigid* (cold) climates, the *temperate* (moderate) climates, and the *tropical* (hot) climates are all part of North America. Because Canada and the United States are such large countries, some of those climates are part of one or both countries.

Directions: Complete the exercise.

1. What is the area of North America? ______________________________

2. What two continents are larger than North America?

   a. ______________________ b. ______________________

3. List four large bodies of water that border North America.

   a. ______________________ c. ______________________

   b. ______________________ d. ______________________

4. Two large countries in North America are the ______________________

   and ______________________ .

5. North America also includes the country of ______________ and the countries

   of ______________________ .

6. List five land features of North America.

   a. ______________________ d. ______________________

   b. ______________________ e. ______________________

   c. ______________________

7. The highest peak in North America is ______________ ______________ at 20,320 feet above sea level.

8. The lowest point in North America is ______________ ______________ at 282 feet below sea level.

9. Frigid means ______________ ; temperate means ______________ ,

   and tropical means ______________ .

10. There are twelve climates on the earth. How many of those twelve are found in

    North America? ______________________________ .

The United States

The United States is the fourth largest country on the earth. The two countries of Canada and the United States make up four-fifths of the continent of North America. The area of the United States is 3,679,245 square miles and has a population of about 240 million. Canada borders the United States to the north, the Atlantic Ocean to the east, the Pacific Ocean to the west, and the Gulf of Mexico and the country of Mexico to the south.

Not only is the United States a very big country, but the location is also very good. Trade can be carried on with South America, Africa, Asia, and Europe. The large oceans also provide safety in times of war.

Land Features of the United States. The United States extends from the Atlantic Ocean on the east to the Pacific Ocean on the west. The distance across the United States is nearly 3,000 miles. From east to west, there are several important land features.

Along the east coast is a lowland region that is narrow in the northeast and gets much wider in the south. The region is called the *Atlantic Coastal Plain*. The lowland plain continues along the Gulf of Mexico. The region covers a large area, as far west as the state of Texas. The region is called the *Gulf Coastal Plain*. The Atlantic and Gulf Coastal Plains are very important in both farming and other industries.

Moving west from the Atlantic Coastal Plain are the *Eastern Highlands*. Sometimes they are called the *Appalachian Mountains*. The highlands, however, are really three separate parts. From east to west they are the *Piedmont Plateau*, the *Appalachian Mountains*, and the *Appalachian Plateau*. The highest parts of the region are in the northern and southern parts. Broad fertile valleys lie within the region and are very good for farming.

West of the Eastern Highlands is a very large region called the *Interior Plains*. The two parts of the Interior Plains are called the *Central Plains* and the *Great Plains*. The dividing line between the Central Plains and the Great Plains is the longitude line marked 100 degrees W.

In the southwestern part of the Central Plains are the *Ozark Highlands*. The Ozark region has two parts, the *Ozark Plateau* and the *Ouachita Mountains*.

The *Rocky Mountains* begin west of the Great Plains. The mountain range runs from Mexico to northern Canada. The Rockies are high mountains, well over 10,000 feet above sea level. The mountains slope down to a large plateau. The plateau covers many states, from Washington State to New Mexico. This region is called the *Intermountain Plateau*.

The far western part of the United States is also a very mountainous region. The *Sierra Nevada* and the *Cascade Mountains* extend from California to Canada. Both mountain ranges are very high. Closer to the west coast is still another mountainous region. The *Pacific Coastal Ranges* follow the coast from California to Canada. Between the Pacific coastal ranges and the Sierra Nevada and Cascade Mountains are large valleys. Some of the best farm lands in the United States are in those valleys.

A. Directions: Match each term in *Column A* with the correct description in *Column B*. Write the letter of your answer on each line.

Column A	Column B
1. Canada ________	a. Eastern border of the United States
2. Atlantic Ocean ________	b. Northern border of the United States
3. about 240,000,000 ________	c. Southern boundary of the United States
4. Gulf of Mexico _______	d. Population of the United States
5. Pacific Ocean ________	e. Western boundary of the United States

B. Directions: Complete the exercise.

1. List the two coastal plains of the eastern part of the United States.

   a. ______________________ b. ______________________

2. What is another name for the Eastern Highlands? ______________________

3. The Eastern Highlands have three parts. What is the name of each part?

   a. ______________________ c. ______________________

   b. ______________________

4. Where is the Interior Plains of the United States located?

   ____

5. What are the two parts of the Interior Plains?

   a. ______________________

   b. ______________________

6. What is the importance of the longitude line marked 100° West?

   ____

7. Where are the Ozark Highlands located?

   ____

8. List four mountain ranges of the western part of the United States.

   a. ______________________ c. ______________________

   b. ______________________ d. ______________________

9. The Intermountain Plateau extends from ________________ to ________________ .

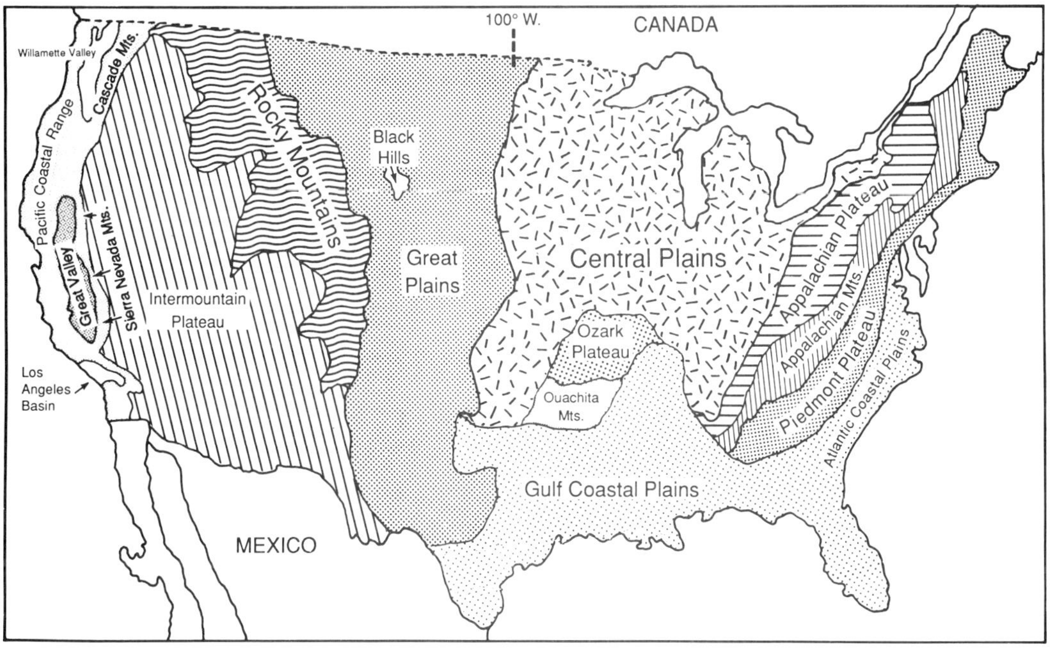

Land Features of the United States

A. Directions: Find each of the land features of the United States on the map above. Write each one in the space provided below.

1. ______________________ 6. ______________________
2. ______________________ 7. ______________________
3. ______________________ 8. ______________________
4. ______________________ 9. ______________________
5. ______________________

B. Directions: Match each boundary of the United States in the right column with its location in the left column. Write the letter or letters of each answer on each line.

1. North ______ a. Canada
2. South ______ b. Mexico
3. East ______ c. Gulf of Mexico
4. West ______ d. Great Lakes

e. Atlantic Ocean

f. Pacific Ocean

Climates of the United States. The United States, with the exceptions of Hawaii and Alaska, lies in the *Temperate Zone*. The southernmost point is about 25 degrees north latitude to the northernmost point of 49 degrees north latitude. The middle latitudes have several climates. The United States has all of those climates and two of the three tropical climates as well. Alaska is in the higher latitudes and adds two more climates to the list. The varied mountain climate is also included. There are highlands in eastern and western United States as well as in Hawaii and Alaska.

The map will help you locate the many climates of our country.

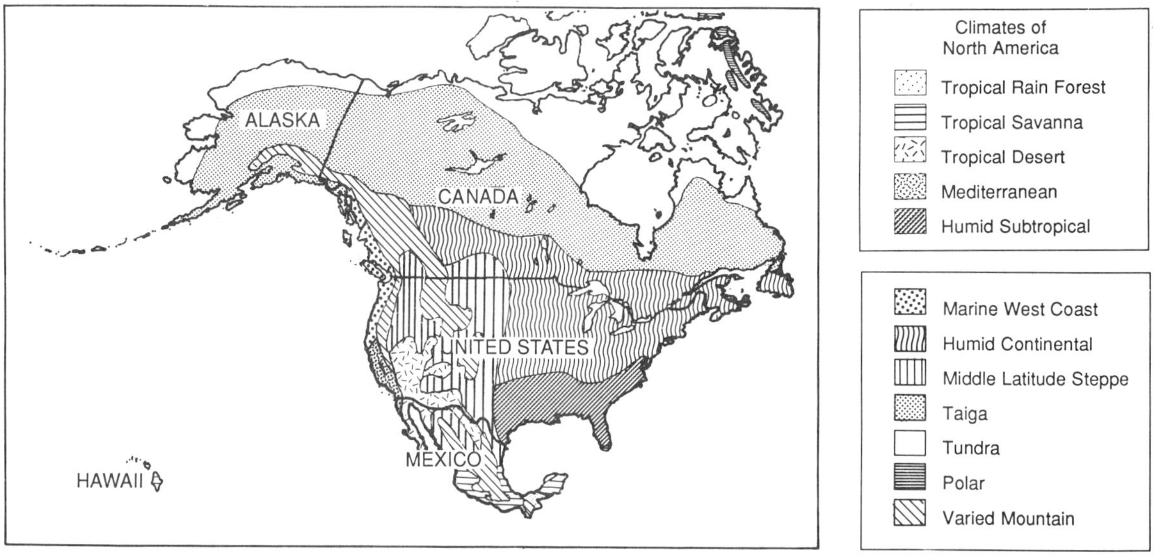

Climates of North America

Directions: Use the map above to complete this exercise.

Tropical climates

a. ______________________ b. ______________________

Middle latitude climates

a. ______________________ d. ______________________

b. ______________________ e. ______________________

c. ______________________

High latitude climates

a. ______________________ b. ______________________

Major Climates of the United States

There are four climatic regions that are well-suited for people to live in. They are the Mediterranean, humid subtropical, marine west coast, and humid continental climates. Most of the people of the United States live in these climatic regions.

The Mediterranean Region. The small Mediterranean region is found only in Southern California. The climate has a hot, dry summer and a mild, wet winter. The summer is long and hot, but the cold California current flows nearby and provides a cool breeze. The daily breezes make the region very comfortable to live in, especially at night. The Mediterranean climate has a water problem. Large water projects which include dams and reservoirs, have been built in California. Water is carried by pipelines to all parts of Southern California. As the population continues to *increase* (get larger) more water will be needed.

The Humid Subtropical Region. Southeastern United States has the humid subtropical climate. The region covers a very large area. Many states are included in this climate. The region extends from eastern Texas to Virginia. Over such a large area, the climate *varies* (is different) from west to east and from south to north. There are two important facts to remember about the climate. To the west, the climate is less wet; and to the north the climate is much colder. The state of Florida has no cold season. Farther north, winter can bring snow and freezing temperatures.

The Marine West Coast Region. Along the west coast of the United States is a small *marine* region. Marine means related to water. Northern California, Oregon, Washington, and western Alaska are in this climatic region. The warm Japanese current flows along the Pacific coast. The westerly winds blow over the warm current. The summer season is mild and wet. Usually the temperature is in the 70's. At times it can get hot, but temperatures in the 90's would be unusual for the region. The winter season is cooler and wet. Snow and freezing temperatures are *seldom* (not often) seen in the marine climate. The climate is comfortable and good to work in.

The Humid Continental Region. Northeastern United States has the humid continental climate. The region is very large. Nearly one-fourth of the United States has this climate. Not only is it large in area, but very large in population as well. More than half of the people of our country live in this climate. The humid continental climate has four seasons—spring, summer, fall (autumn), and winter. Each of the seasons brings a change to the region. Spring and fall are very pleasant seasons. The summer season can get very hot and very humid. At times, the season can be very uncomfortable. The winter season often is cold and snowy. In some parts of the region, the low temperature can be far below 0° F. In spite of these extremes, the climate has a very large population.

Other Climates of the United States. The tropical desert, tropical savanna, middle latitude steppe, taiga, tundra, and varied mountain are other climates found in the United States. The tropical desert and middle latitude steppe are very dry. The taiga and the tundra of Alaska are very cold. The tropical savanna regions of Hawaii and southern Florida are small in area. The climate, however, is warm all year long and does attract more people. In the varied mountain climate, people live in the valleys and the lower parts of the mountains. The higher lands are just too cold and unsuited to attract a large population.

A. Directions: Complete the exercise.

1. Where is the humid continental climatic region of the United States located?

____

2. List three facts that describe the humid continental region of the United States.

   a. ______________________________

   b. ______________________________

   c. ______________________________

3. List the six other climates of the United States.

   a. ______________ d. ______________

   b. ______________ e. ______________

   c. ______________ f. ______________

4. Of the climates of the United States, which one do you live in?

____

____

B. Directions: Complete the exercise.

1. Most of the people in the United States live in four climatic regions. List those climatic regions.

   a. ______________________ c. ______________________

   b. ______________________ d. ______________________

2. List three facts that describe the Mediterranean climate of Southern California.

   a. ______________________

   b. ______________________

   c. ______________________

3. Where is the humid subtropical climatic region of the United States located?

   ____

4. List three facts that describe the humid subtropical climate.

   a. ______________________

   b. ______________________

   c. ______________________

5. Where is the marine west coast climatic region of the United States located?

   ____

6. List three facts that describe the marine west coast climatic region.

   a. ______________________

   b. ______________________

   c. ______________________

The Regions of the United States

The United States is made up of 50 states. Forty-eight of those states extend from the Atlantic Ocean to the Pacific Ocean and are divided into separate regions. Alaska and Hawaii do not *border* (touch) any other of the 48 states. Alaska borders the country of Canada, and Hawaii is a group of islands in the Pacific Ocean. In 1959, Alaska and Hawaii became states.

Directions: The forty-eight states that touch one another are divided into separate regions. Use the map below to locate the regions. Write the names of the four regions below.

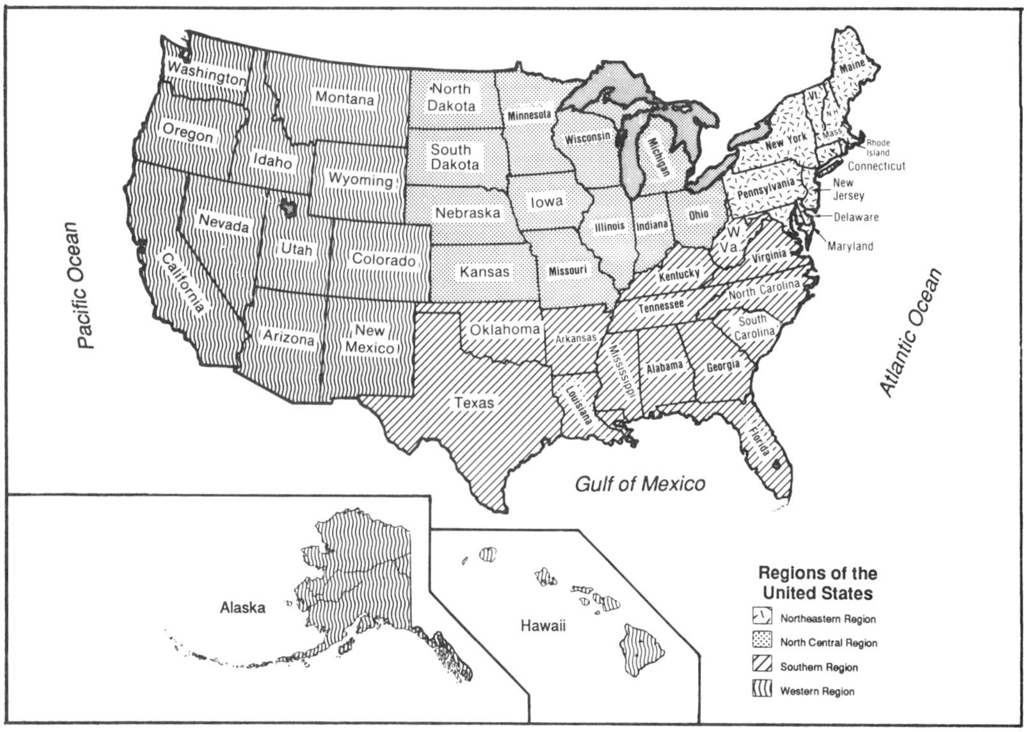

Regions of the United States

1. a. ______________________ c. ______________________

   b. ______________________ d. ______________________

2. What water body is to the east of the United States? ______________________

   To the south? ______________________

   To the west? ______________________

The Northeastern Region

The New England States. The Northeastern region is usually divided into two separate regions. The six New England states make up one region and the five Middle Atlantic states make up the second region. The New England states are Maine, Vermont, New Hampshire, Connecticut, Massachusetts, and Rhode Island. The Middle Atlantic states are New York, Pennsylvania, New Jersey, Delaware, and Maryland.

Industries of the Northeastern States. The Northeastern region of the United States has a great variety of industries. New England became a manufacturing center in the early colonial period. Machinery, *textiles* (cloth), shoes, paper products, and tools are some of the products made in New England. In recent years, electronic equipment has become an important industry. Marble and granite are mined. Tobacco, maple trees for syrup, hay, and cranberries are important *cash* (money) crops. In addition, New England has many good harbors which are well-suited for shipbuilding, fishing, and trade. Many people live in the towns and cities.

Directions: Complete the exercise.

1. List the six New England states. Then, write the capital of each state. Refer to the map on page 82.

State	Capital
a. ____________	____________
b. ____________	____________
c. ____________	____________
d. ____________	____________
e. ____________	____________
f. ____________	____________

2. List five manufactured products made in New England.

a. ____________ d. ____________
b. ____________ e. ____________
c. ____________

3. List four *agricultural* (farm) products grown in New England.

a. ____________ c. ____________
b. ____________ d. ____________

4. List three other industries of New England.

a. ____________ c. ____________
b. ____________

Map of the Regions of the United States

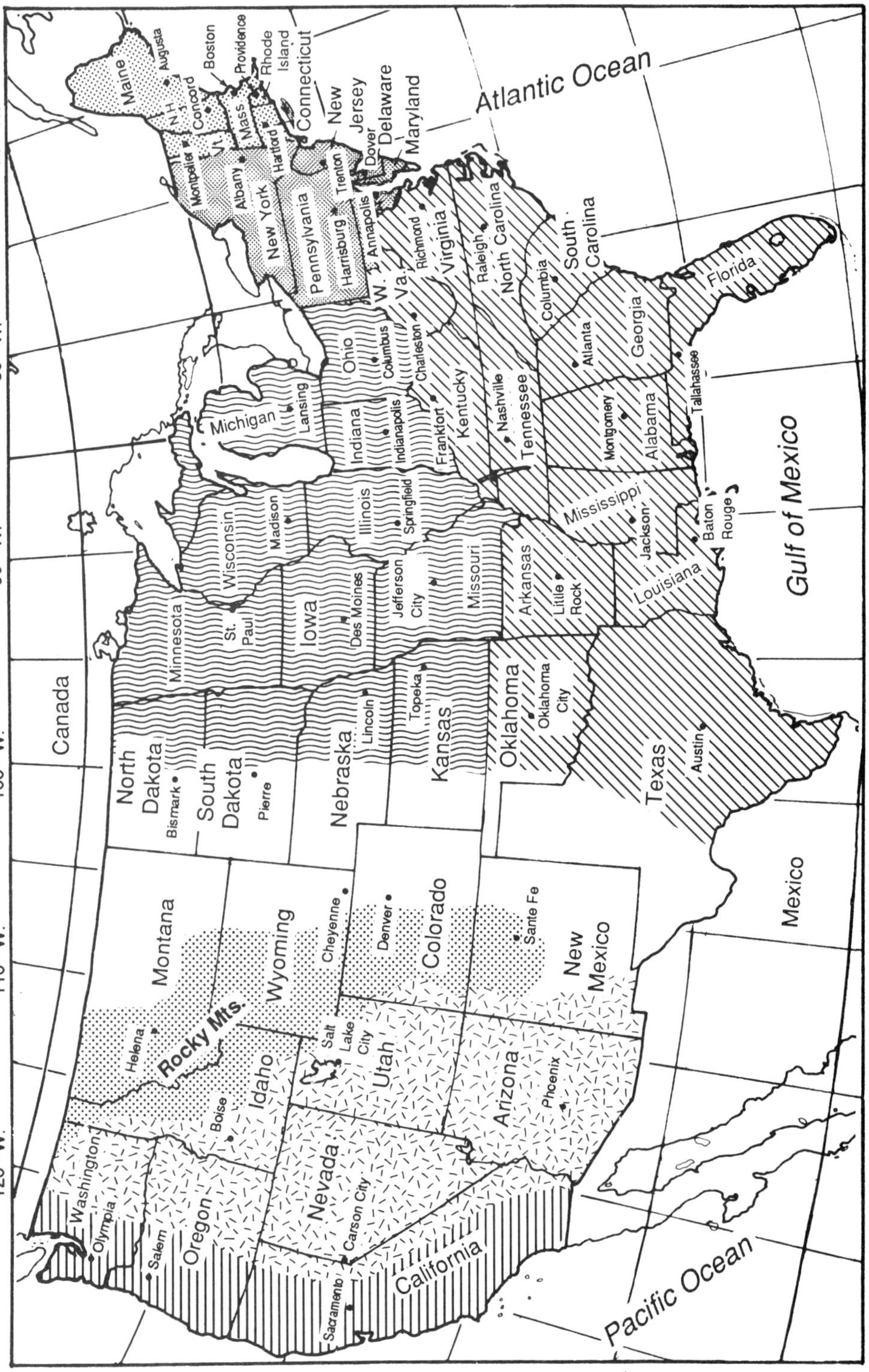

The Middle Atlantic States. New York, Pennsylvania, New Jersey, Delaware, and Maryland are the states of the Middle Atlantic region. The location of these states is of great importance. Much of the trade of our country goes through this region. New York City, New York; Philadelphia, Pennsylvania; and Baltimore, Maryland are very large ports and important trade centers. Railroads, excellent roads, and many airlines connect the region to the south, north, and east. This transportation network makes the region very important to other parts of the country.

The Middle Atlantic states have many industries. The farmlands and climate are well-suited for many crops. Wheat, corn, fruits, and vegetables are grown throughout the region. *Poultry* (chickens, turkeys, etc.) and *livestock* (cattle, hogs, etc.) are also important farm products. Many products are also manufactured. Coal mines, shipbuilding, steel mills, chemicals, and auto assembly plants are located in the region. The Middle Atlantic states are among our country's most productive states.

Directions: Complete the exercise.

1. List the five Middle Atlantic states. Then, write the capital of each state. Refer to the map on page 82.

	State	**Capital**
a.	__________	__________
b.	__________	__________
c.	__________	__________
d.	__________	__________
e.	__________	__________

2. Why is the location of the Middle Atlantic states very good for trade?

____

3. List four farm crops produced in the Middle Atlantic states.

   a. __________ c. __________

   b. __________ d. __________

4. List four other industries that are found in the Middle Atlantic states.

   a. __________ c. __________

   b. __________ d. __________

The North Central Region

The North Central Region is sometimes called the *Midwest*. There are twelve states in this region. They are Ohio, Michigan, Indiana, Illinois, Wisconsin, Minnesota, Iowa, Missouri, and the eastern parts of Kansas, Nebraska, South Dakota and North Dakota. Several states of the region border along the Great Lakes. Many important cities are located along the lakes and are important industrial and trade centers. Two large rivers also flow through the North Central states. Several cities are located along the Missouri and Mississippi Rivers.

Farming in the North Central States. Farming is a very big industry in this region. Spring and winter wheat are grown in the drier western part of the Midwest. Corn is grown in the wetter eastern part of the region. Minnesota and Wisconsin are leading dairy states. Animals such as cattle and hogs are raised in large numbers. The region is sometimes called the "breadbasket of our country."

The Industrial North Central States. The North Central states, or Midwest, has many advantages for heavy industries. Heavy industries use coal and iron ore. The iron ore is made into steel and used to make many products. Large cities such as Chicago, Illinois; Gary, Indiana; Detroit Michigan; and Cleveland, Ohio are located along the shores of the Great Lakes. The raw materials needed to make products are shipped on the Great Lakes to these big cities. Railroads and super highways are also used to ship raw materials and finished products. Many important products are made in the region. Automobiles, farm equipment, machinery, airplane parts, home appliances, and furniture are just a few of the products made in the region. In addition, farm crops are also made into products. Farming and manufacturing have made the North Central states our leading industrial region.

Directions: Complete the exercise.

List the 12 North Central states. Refer to the map on page 82.

1. ______________________
2. ______________________
3. ______________________
4. ______________________
5. ______________________
6. ______________________
7. ______________________
8. ______________________
9. ______________________
10. ______________________
11. ______________________
12. ______________________

The Southern Region

The American southland is a very large region. There are fourteen states included in this region. The states are Virginia, West Virginia, North Carolina, South Carolina, Georgia, Florida, Alabama, Mississippi, Louisiana, Arkansas, Tennessee, Kentucky, and the eastern parts of Texas and Oklahoma. The region is made up of a very large lowland plain that extends very far inland. Some higher lands, called the *Oauchita Mountains* and the *Ozark Plateau,* are found in Missouri, Arkansas, and Oklahoma. The Appalachian Mountains run through the western part of the region from West Virginia to Alabama. Many of the states border along the Atlantic Ocean and the Gulf of Mexico. The long coastline and the rivers are very useful for *commerce* (trade).

Farming in the Southern Region. The southland has three important advantages for farming. The climate is warm and wet, and the land is level and fertile. Because the region is so large, the climate does vary to some extent. The southern part is warmer and has very little cold weather. The northern part is much colder during the winter months. A great variety of crops are grown. Cotton, tobacco, corn, peanuts, rice, sugar cane, citrus fruit, and vegetables are some of the special crops of the region. Trees also are grown as a crop. The trees are used for paper, lumber, building products, and furniture. New machines and better methods have made the southland an outstanding farming region.

The New Industries of the South. New industries have developed in the South. The region no longer depends just on farming. *Textiles* (cloth), *synthetic* (man-made) fibers, petroleum, iron and steel mining, food processing, and fishing are now major industries. There are many harbors along the Gulf of Mexico and the Atlantic Ocean. Rivers such as the Mississippi, the Ohio, and the Arkansas provide good water transportation, as well as water power. New and modern roads have been built; railroads have been improved and air travel has increased. The population of the region has increased very quickly. Dallas and Houston, Texas; New Orleans, Louisiana; Mobile and Birmingham, Alabama; Tampa, Florida; Atlanta, Georgia; Nashville, Tennessee; and Richmond, Virginia are some of the big cities of the region.

Directions: List the 14 Southern states. You may refer to the map on page 82.

1. ______________________
2. ______________________
3. ______________________
4. ______________________
5. ______________________
6. ______________________
7. ______________________
8. ______________________
9. ______________________
10. ______________________
11. ______________________
12. ______________________
13. ______________________
14. ______________________

The Western Region

The Great Plains States. The Great Plains is a large region that is located between the Central Plains to the east and the Rocky Mountains to the west. The states that are a part of the Great Plains are the western parts of Texas, Oklahoma, Kansas, Nebraska, North Dakota, South Dakota, and the eastern parts of Montana, Wyoming, Colorado, and New Mexico. The 100° W. longitude line is an easy way to tell where the Great Plains begin. All the land between the line 100° W. and westward to the Rocky Mountains is the Great Plains. The states of the Rocky Mountain Region are a part of the Intermountain and the Great Plains states.

A Semiarid Region. *Semiarid* means that precipitation averages between 10-20 inches a year. The region is dry because it lies east of the high Rocky Mountains. Most of the *precipitation* (rain, snow, etc.) falls before the winds get to the Great Plains. Therefore, the Great Plains is not very good for farming.

Farming in the Great Plains. Sheep and cattle ranches are very large in the Great Plains. A small ranch can be about 1,000 acres. A large ranch can be about 50,000 acres to 100,000 acres. Ranches are big because there is only a small amount of grass for the animals to *graze* (feed) on. Along the wetter part, near the 100° W. longitude line, wheat is grown. The lack of precipitation causes crop failure and hard times for the farmers. The region has a small population. The cities are small and widely scattered. The beautiful scenery of the region attracts a large number of tourists each year. Hotels, motels, cabins, and campsites provide jobs for many people of the Great Plains.

Directions: Complete the exercise.

1. List the Great Plains states. Refer to the map on page 82.

   a. ____________ f. ____________
   b. ____________ g. ____________
   c. ____________ h. ____________
   d. ____________ i. ____________
   e. ____________ j. ____________

2. Where is the Great Plains located? ____________

   ____

3. Why is it very hard to farm in the Great Plains? ____________

   ____

4. Write two facts about ranches in the Great Plains.

   a. ____________
   b. ____________

The Intermountain States. Between the Rocky Mountains to the east and the Sierra Nevada and Cascade Mountains to the west is a large plateau. The large plateau contains the Intermountain states. Parts of the states of Washington, Oregon, California, Idaho, Utah, New Mexico, Arizona, and Nevada are in this region. Arizona and Nevada are completely within this region.

Industries of the Intermountain States. Like the Great Plains, the Intermountain region is very dry. Wheat, sugar beets, potatoes, and fruits are grown in areas that have water for irrigation. The major industries, however, are mining, lumbering, and tourism. Many minerals are found in this region. The important ones are copper, gold, silver, uranium, zinc, and lead. Each year millions of tourists visit the Intermountain states to see the beautiful Grand Canyon, Bryce Canyon, the Hoover Dam, and Yellowstone National Park.

Although the population is not as large as some of the other regions of the United States, there are some major cities. Some of these important cities are Phoenix, Arizona; Albuquerque, New Mexico; Denver, Colorado; Salt Lake City, Utah; Boise, Idaho; and Spokane, Washington.

Directions: Complete the exercise.

1. List the eight Intermountain states. Refer to the map on page 82. You will see that only parts of the states belong to that region.

   a. ______________________ e. ______________________
   b. ______________________ f. ______________________
   c. ______________________ g. ______________________
   d. ______________________ h. ______________________

2. Where are the Intermountain states located? ______________

   ____

3. List four crops grown in the Intermountain states.

   a. ______________________ c. ______________________
   b. ______________________ d. ______________________

4. List three other important industries of the Intermountain states.

   a. ______________________ c. ______________________
   b. ______________________

5. List four places that people like to visit in the Intermountain states.

   a. ______________________ c. ______________________
   b. ______________________ d. ______________________

The Pacific Coastal States. Along the west coast of the United States are the Pacific Coastal states. The region is located west of the Cascade Mountains in the north and the Sierra Nevada Mountains in the south. Parts of the states of Washington, Oregon, and California are in this region. A narrow coastal plain, two major valleys, and a coastal range of mountains make up this region.

Climates of the Pacific Coastal States. Northern California, Oregon and Washington have the marine west coast climate. The summer season is mild and wet. The winter season is cool and also wet. Southern California has the Mediterranean climate along the coast and the tropical desert climate farther inland.

Farming in the Pacific Coastal States. Farming in Oregon and Washington is a major industry. The Williamette Valley and the Puget Sound lowlands are very good for cool weather crops. Some of the farm products of the region are hay, livestock, vegetables, and fruits, especially apples. Farther south in the Mediterranean climate, many different crops are grown. The hot climate is good for cotton, wheat, corn, grapes, and citrus fruit, as well as other crops. Because there is almost no rain in the summer season, much of the land is irrigated. Irrigation is also needed in the desert region due to the dry climate. The desert region has 0-10 inches of rainfall a year. Large irrigation projects have been built to make farming possible. The Imperial Valley of California is one of the best farming regions in the world because of irrigation.

Other Industries of the Pacific Coastal States. There are many important industries in the Pacific Coastal states. The giant Sequoia and Douglas fir trees grow well in this region. Lumbering is a very big business. Fishing is also important. Salmon, tuna, sardines, and many other fish are caught and sent to *canneries.* Canneries are places that clean and package the fish for shipment. Mining is done in California and Washington. Gold, silver, copper and petroleum are the major minerals of the region. Many important products are made there such as airplanes, ships, electronic equipment, automobiles, machinery, and paper. California, which is the center of the *motion picture* (movie) industry, also has a very large tourist industry. Millions of people like to visit the region to see the beautiful scenery. California has the largest population of all the states in our country. Nearly 27 million people live there. Some of the important cities of the Pacific Coastal Region are San Diego, Los Angeles, and San Francisco in California; Portland, and Salem, Oregon; and Seattle in Washington.

Directions: Complete the exercise.

1. List the three Pacific Coastal Region states.

   a. ______________________ c. ______________________

   b. ______________________

2. List the three climates of the Pacific Coastal states.

   a. ______________________ c. ______________________

   b. ______________________

3. Why does farming depend on irrigation in the Mediterranean and tropical desert climates? ____

   ____

4. List four important industries of the Pacific Coastal states. Write one fact about each.

   a. ____

   b. ____

   c. ____

   d. ____

5. List six major cities of the Pacific Coastal states. Include the state of each city.

   a. ____

   b. ____

   c. ____

   d. ____

   e. ____

   f. ____

Alaska our 49th State

Alaska is our largest state. The area is about 589,757 square miles and has a population of about 534,000. It is located in the northwestern corner of North America. The country of Canada borders Alaska to the east. Across the Bering Strait is the country of the Soviet Union. Alaska was bought in 1867 from *Russia* (Soviet Union). In 1959, Alaska was added as our 49th state.

Farming in Alaska. Much of Alaska is very mountainous land. Mt. McKinley, located in Alaska, is the highest mountain in North America. It stands 20,320 feet above sea level.

The southwestern part has a marine west coast climate and is mild and wet. Farther inland and northward the climate is much colder. Many of the Eskimo population live in this colder tundra region. The most populated part of Alaska is along the southern and southeastern coasts. Farming is carried on in the Matanuska Valley, which is near the city of Anchorage. The summer is mild and wet and the days are very long. Many crops are grown in this fertile valley.

Alaska Today. The population of Alaska is beginning to increase very rapidly. The *Alcan* (Alaska-Canada) highway connects Fairbanks, Alaska to southern Canada. The new road makes traveling in Alaska much easier and faster. The Alaskan pipeline carries oil from northern Alaska to southern Alaska. Many people went to Alaska to live and work on the pipeline. Alaska also has a growing tourist industry. No longer is Alaska called the "ice box." The scenery is beautiful. Camping and fishing are enjoyed by many visitors. The major cities of Alaska are Anchorage, Fairbanks, Sitka, Ketchikan, and Juneau. The capital of Alaska is Juneau.

Directions: Fill in each blank with a term from the box.

1867	ice box	Alcan	Anchorage
534,000	Fairbanks	Ketchikan	largest
1959	Sitka	Mt. McKinley	colder
Juneau	oil	Russia	southeastern
Matanuska			

Alaska is our ____________________ state. The population of Alaska is about ________________. The United States bought Alaska from ________________ in __________. In __________, Alaska became our 49th state. Much of Alaska is very mountainous. The highest peak is____________________.

The Eskimo people live in the ________________ part of Alaska. Most of the cities are along the southern and ____________________ coasts of Alaska. Farming is carried on in the ________________ Valley, which is near the city of ________________. Other important cities of Alaska are ________________, ________________, ________________, and ________________.

With the new________________ highway and the discovery of _______, Alaska can no longer be called the __________ __________.

Hawaii

Hawaii first became a territory of the United States in 1898. In 1959, Hawaii was added to our country as the 50th state. The volcanic islands are located in the Pacific Ocean about 2,400 miles west of California. There are many islands, but not all of the islands have people living on them. The total area of all of the islands is 6,450 square miles, and the population is over one million. The capital city is Honolulu (*Hahn uh LOO loo*).

Industries of Hawaii. The biggest and most important industry in Hawaii is *tourism* (visitors). Millions of people visit the islands to enjoy the nearly perfect climate, beautiful scenery, white sandy beaches, native Hawaiian food and entertainment. The naval base at Pearl Harbor on the island of Oahu (*Uh wah hoo*) is also important. The military base provides many jobs for the people of the island. Sugar cane and pineapples are the two special crops of the islands. Both crops are grown on large plantations. Manufacturing is increasing, and a greater variety of crops are being grown. These changes provide more jobs for the people of Hawaii.

Directions: Write a summary paragraph about Hawaii in which you use each of the items listed below. The items may be used in any order. Underline each of the items. Use your own paper.

visitors	50th state	1898
Pearl Harbor	6,450 sq. miles	territory
1959	volcanic	pineapples
sugar cane	Honolulu	one million

Directions: Complete the exercise below.

1. How does the United States compare in size with the other countries of the earth?
____

2. What is the area of the United States? ______________________
What is the population? ______________________

3. What is the boundary of the United States on the:
a. north ______________ c. east ______________
b. south ______________ d. west ______________

4. List the two large coastal plains of the United States.
a. ______________ b. ______________

5. List the two large Interior (inland) Plains of the United States.
a. ______________ b. ______________

6. List the three regions that make up the Eastern Highlands.
a. ______________
b. ______________
c. ______________

7. List the two regions of the Ozark Highlands.
a. ______________ b. ______________

8. List the four mountain ranges of the western region of the United States.
a. ______________ c. ______________
b. ______________ d. ______________

9. How many climates does the United States have? ______________

10. List the four climates of the United States that have the largest populations.
a. ______________ c. ______________
b. ______________ d. ______________

11. The states of the United States are divided into regions. List the names of those regions.
a. ______________ e. ______________
b. ______________ f. ______________
c. ______________ g. ______________
d. ______________

12. Both Alaska and Hawaii became states in the year __________ .

Answers

Page 7

A. 1. March 21, June 22, September 23, December 22
2. June 22
3. December 22
4. spring, March 21 and fall, September 22
5. Tropic of Cancer
6. Tropic of Capricorn

B. 1. round 3. 8,000 5. axis 7. day, night
2. globe 4. 25,000 6. rotation, revolution 8. seasons

Page 8

The earth is made up of several landforms such as *mountains*, *plateaus*, *hills*, and *plains*. The surface of the earth is not smooth and has many *rivers* which cut through to the coastal plains. Rivers are very important because they provide a supply of fresh *water*.

The earth's surface is about *70* percent water and *30* percent land. The great oceans are the *Atlantic*, *Pacific* , *Indian*, and the *Arctic*. Lately, *6* continents are more often identified. They are Eurasia, *South America*, *North America*, *Africa*, *Antarctica*, and *Australia*.

Page 10

A. 1. Torrid, North Temperate South Temperate, North Frigid, South Frigid
2. Cancer, Capricorn
3. Cancer, Arctic
4. Capricorn, Antarctic
5. Arctic, North Pole
6. Antarctic, South Pole
7. equator

B. 1. B 3. D 5. I 7. E 9. H
2. A 4. C 6. F 8. G 10. J

Page 12

A. 1. True 3 False 5. True 7. True 9. False
2. True 4. True 6. True 8. True 10. True

B. 1. prime meridian 4. 15 7. 24
2. 180 degrees 5. 24 8. 5 a.m.
3. 4 6. Greenwich 9. 3 p.m.

Page 14

1. farmers fishermen business people airline pilots
2. It affects people and at times can be a source of difficulty.
3. The daily condition of the atmosphere at a particular time.
4. A blanket of gases that surround the earth.
5. a. air pressure e. precipitation
   b. humidity f. clouds
   c. temperature g. sunshine
   d. wind velocity and direction
6. 14.7 pounds per square inch at sea level
7. We adjust to it.
8. barometer
   As air pressure increases, the mercury goes up. As air pressure decreases, the mercury goes down.
9. high
10. low
11. lower

Pages 15 and 16

A. 1. gas 5. humidity 9. more
2. wate 6. 100 percent 10. down
3. liquid to a gas 7. saturated
4. oceans 8. temperature

B. 1. Daily condition of the atmosphere.
2. Gaseous blanket around the earth.
3. The weight of air.
4. Instrument that measures air pressure.
5. Over 30 inches at sea level on a barometer.
6. Under 30 inches at sea level on a barometer.
7. Water vapor in the air.
8. Gas – water in a gaseous condition.
9. 100 percent (full).
10. Refers to the amount of water vapor in the air in relation to the temperature.

Page 18

Temperature is the *warmth* and *coolness* of the atmosphere. The instrument that measures temperature is called a *thermometer*. The measurement of temperature used in the United States is made on the *Fahrenheit* scale. The boiling point of water is *212* degrees, and the freezing point is *32* degrees. Another measurement used in all other countries is the *Celsius* or *centigrade* scale. The boiling point of water on that scale is *100* degrees, and the freezing point of water is *0* degrees.

When air is heated, it *expands*; when air is cooled, it *contracts*. This causes the upward and downward movement of air which is called *convection*.

A high pressure air mass shows a reading on a barometer of over *30 inches*. A low pressure air mass shows a reading of under *30 inches* on a barometer. The difference in air pressure causes air to have motion called *wind*.

Page 20

A. 1. Evaporation
2. Condensation
3. Clouds
4. Precipitation

B. 1. a. oceans
   b. lakes
   c. rivers
   d. vegetation
   e. ground water
2. gas
3. liquid
4. water
5. precipitation

Page 21

1. a. convectional b. cyclonic
2. thunderstorms
3. Middle latitudes
4. a. Lasts a long period of time.
   b. Takes place over a large area.
   c. There is little actual precipitation.

Page 22

1. a. pressure e. humidity
   b. temperature f. clouds
   c. wind velocity and direction g. sunshine
   d. precipitation
2. Answers will vary.
3. a. The process by which the earth absorbs heat.
   b. The process by which the earth gives off warmth or coldness.
   c. The upward and downward movement of warmed and cooled air.

Page 24

1. Western Australia 3. New South Wales 5. 926 square miles
2. Tasmania 4. Tasmania

Pages 25 and 26

A. 1. Aborigines 5. Aussies 8. iron ore, lead, gold, copper, coal, oil, bauxite
2. bushmen 6. non-urban 9. city
3. little, far 7. steel mills, automobile assembly plants, flour mills 10. 85
4. yourself

B.

	State	*Capital*
1.	a. Tasmania	Hobart
	b. Victoria	Melbourne
	c. South Australia	Adelaide
	d. Queensland	Brisbane
	e. Western Australia	Perth
	f. New South Wales	Sidney

	City	*State*
2.	Sidney	New South Wales
	Melbourne	Victoria

Page 28

1. Rio Grande, South America
2. a. Mexico
   b. Carribean Sea
   c. Central America
   d. South America
3. Romans
4. a. Christopher Columbus
   b. Balboa
   c. Pizarro
   d. Cortes
5. Portugal
6. Indians
7. independent

Page 29

1. Topography
2. South America
3. 10 degrees, 55 degrees
4. a. Venezuela
   b. Colombia
   c. Ecuador
   d. Peru
   e. Bolivia
   f. Chile
   g. western Argentina
5. a. Pampas
   b. Amazonian Basin
6. land bridge, South America
7. mountains, valleys, coastal plains, plateaus
8. seven
9. plateau, mountains
10. climate

Pages 30 and 31

1. The backbone of a continent. An example is the Andes mountains in South America.
2. Grasslands of Argentina.
3. Table top land 4,000 - 8,000 feet.
4. A large farm like a plantation.
5. Another name for land features (what the land looks like).

Page 32

1. Brazil
2. South America
3. Soviet Union, Canada, China, United States
4. half
5. land features, climate
6. Amazon River Basin
7. 3,900, largest
8. plateau, Brazilian, coastal plain

Page 33

1. equator, Tropic of Capricorn
2. tropical rain forest
3. trees, rainfall
4. 40 percent
5. Indians
6. tropical savanna, humid subtropical
7. humid subtropical

Page 34

1. Amazon River
2. latex
3. latex
4. boom
5. boom, important
6. Brasilia
7. 600
8. Brazil
9. Rio de Janeiro

Page 35

1. Accept any reasonable answer.
2. a. shortage of jobs
   b. low pay for workers
   c. over-crowded cities
   d. fast growing population
   e. poor railroad and roads
3. Pay back is difficult unless growth improves. Bad debts could cause more loans and loss of confidence in Brazil's capacity to make progress.

Page 37

Argentina is the *second* largest country in *South America*. The area is *1,065,200* square miles with a population of about *31* million. Only *Brazil* and *Mexico* have a larger population in Latin America.

The major land features of Argentina are the Western *Mountains*, a large plain called the *Pampa*, and a windswept region far to the south called *Patagonia*.

The climates of Argentina are very much like those of the *United States*. The most productive region has a *humid subtropical* climate. Summers are *hot*; winters are *mild*, and *rainfall* is year round. Not only is the climate good for farming, but the region is very large and extends nearly *400* miles inland.

Page 39

1. Pampa
2. Buenos Aires
3. a. cattle b. corn c. wheat d. dairy products e. fruit
4. a. auto assembly
   b. flour mills
   c. meat processing plants
   d. textiles
   e. chemical plants
5. a. skilled workers b. valuable minerals c. rich farm lands
6. Accept any reasonable answer.

Page 41

1. largest, population
2. Eurasia
3. equator, Arctic Circle
4. climates
5. Himalayas
6. Mount Everest, 29,000
7. a. tropical rain forest
   tropical drylands
   tropical savanna
   b. middle latitude drylands
   humid continental
   humid subtropical
   c. taiga tundra

Page 42

Accept any reasonable answer.

Page 43

Japan is a small *island* country in southeastern *Asia*. There are *4* major islands, the largest of which is *Honshu*. Most of the land of Japan is made up of *volcanic* mountains. The highest peak in Japan is *Mount Fuji*, at *12,388* feet above sea level.

Japan extends from about *31* degrees north latitude to *46* degrees north latitude. The latitude of Japan when compared to our country would be from *Maine* to southern *Georgia*. The three most southern islands of Japan are *Honshu*, *Shikoku*, and *Kyushu*. The most northern island is called *Hokkaido*. Southern Japan has the *humid subtropical* climate. Northern Japan has the *humid continental* climate.

Page 45

1. Japan was opened to the Western ideas.
2. It took Japan 50 years to change to an industrial country.
3. Japan was defeated by the United States. Their empire was destroyed.
4. Farming is difficult in Japan because of mountainous land; it is a small country, and the population is very large.
5. Terracing is done by notching out level pieces of land. Step-like sections of level land extend up the hillside.
6. Fishing provides a source of food supply.
7. Rapid progress was made because Japan used already known technology.
8. a. cameras
   b. automobiles
   c. steel
   d. electrical products
   e. shipbuilding
   f. textiles
   g. electronic equipment
9. Japan is the third largest industrial nation in the world.
10. a. pollution b. crowded cities c. need for trade

Page 46

1. small
2. Australia
3. 4,063,000
4. 3,615,000
5. population
6. North America, South America
7. English
8. Europeans

Page 48

	Peninsula	*Country or Countries*
a.	Scandinavian	Norway, Sweden
b.	Jutland	Denmark
c.	Iberian	Spain
d.	Italian	Italy
e.	Balkan	Albania, Greece, Yugoslavia, Bulgaria, Turkey

Page 49

A. 1. Marine west coast climate – cool, rainy summers and mild, rainy winters.
   2. Taiga and tundra climates – cold winters, little precipitation; summers are short; moderate to cool temperature.
   3. Mediterranean climate – dry, hot summers; mild and wet winters.

B. 1. 221,207 square miles; 56,343,000
   2. It is the third largest.
   3. a. Mediterranean Sea
      b. Bay of Biscay, Atlantic Ocean, the English Channel
   4. The location of France is excellent for trade because of its access to bodies of water.

Page 51

A. 1. True 2. True 3. True 4. True 5. True

B. The *Pyrenees* Mountains separate *France* from *Spain*. The *Alps* separate France from *Switzerland* and *Italy*. Three important advantages that France has are an excellent *location*, a large lowland *plain*, and excellent *rivers*.

C. 1. Atlantic Ocean 2. Bay of Biscay 3. Mediterranean Sea

D. 1. Marine west coast – Winters are mild and there is light rain. Summers are pleasant and rainfall is frequent.
   2. Mediterranean – Summers are hot and very dry. The winter is mild and rainy.

Page 53

A. 1. large
   2. 30-40 acres
   3. more
   4. many
   5. important
   6. Paris

B. 1. airplanes
   2. railroad equipment
   3. aluminum products
   4. farm equipment
   5. automobiles
   6. fine silks

C. 1. Le Havre
   2. Bordeaux
   3. Toulouse
   4. Marseilles
   5. Lyon
   6. Cherbourg

D. The important city on the island of Corsica is *Ajaccio*. It was the birthplace of *Napoleon Bonaparte*.

Page 54

1. 11,707,000 square miles; over 600 million
2. Second in area; third in population
3. a. Europe b. Asia
4. Vasco da Gama late 15th century
5. Little was known of the interior of Africa.

Page 56

Most of Africa is made up of a very large *plateau*. The *Atlas* Mountains are found in *northwestern* Africa. The mountains are very close to the coasts of the *Atlantic* Ocean and the *Mediterranean* Sea. The plateau drops sharply to the coast. The coastal plain is very *narrow*, and the coastline has very few good *harbors*. Important rivers that cross Africa are the *Nile*, the *Niger*, the *Congo*, the *Zambezi*, and the *Orange*. The Nile River is the *longest* river in the world. It is over *4,100* miles long. Along those great rivers are many *waterfalls*. The waterfalls make it hard for *ships* to travel on the rivers. The only lowland regions are found in *western* and *northeastern* Africa.

Page 57

1. a. equator 0 degrees
   b. Tropic of Cancer 23 1/2 degrees N.
   c. Tropic of Capricorn 23 1/2 degrees S.
2.

	Climate	***Location***	***Seasons***
a.	*tropical rain forest*	*central*	*hot, wet*
b.	*tropical savanna*	*north and south of tropical rain forest*	*hot & dry; hot & wet*
c.	*tropical desert*	*north and south of savanna*	*hot & dry; cooler & dry*
d.	*Mediterranean*	*northern and south-western Africa*	*hot & dry; mild & wet*
e.	*humid subtropical*	*southeastern Africa*	*hot & wet; mild & wet*

Page 57 (continued)

3. a. Sahara b. Kalahari
4. Grasslands
5. The summers are dry; therefore, there is a lack of water.
6. The humid subtropical

Page 59

1. One part is north of the Sahara Desert and borders along the Atlantic Ocean and the Mediterranean Sea. The other is south of the Sahara and is called sub-Sahara Africa.
2. The Sahara Desert
3. In western Africa near the equator.
4. 92,000 square miles; about 13 million
5. a. Ivory Coast b. Togo c. Burkina Faso
6. Accra
7. a. Tropical rain forest – has one season which is hot and wet.
   b. Tropical savanna – has two separate seasons. Half of the year is hot and very wet. The other half is warm and nearly dry.

Page 60

Ghana was once called the *Gold Coast*. For over a hundred years, Ghana was a colony controlled (ruled) by the *British*. In the year *1957*, Ghana became an independent country.

The most important industry of Ghana is *farming*. The important crop is *cacao* which is used to make *chocolate*. Some other important crops are *cotton*, *coffee*, *latex*, and *palm oil*.

A large dam was built on the *Volta* River. During the dry season, water from the lake behind the dam is used to *irrigate* the land.

Ghana also has some very important minerals. *Manganese* is used to make steel, and *bauxite* is used to make aluminum. *Gold* and *diamonds* are also sold to other countries.

Ghana could be the leading country of Western Africa. The big problem has been the frequent changes in the *government*.

Page 61

Country	*City*	*Desert*
Togo	Accra	Sahara
Ivory Coast		Kalahari
Burkina Faso		
Ghana		
River	***Water Body***	***Mountain***
Congo	Mediterranean Sea	Atlas
Nile	Atlantic Ocean	
Zambezi		
Orange		
Volta		

Page 62

Paragraphs will vary.

Page 64

1. 1) Turkey
   2) Syria
   3) Lebanon
   4) Israel
   5) Jordan
   6) Cyprus
   7) Iraq
   8) Iran
   9) Egypt
   10) Saudi Arabia
   11) Yemen Arab Republic
   12) People's Democratic Republic of Yemen
   13) Oman
   14) United Arab Emirates
   15) Qatar
   16) Bahrain
   17) Kuwait
2. It is located at the far eastern end of the Mediterranean Sea. Europe and Africa border the Middle East to the northwest and southwest.
3. In earlier times, the location of the Middle East was the crossroads of land and water travel. Ships of many countries use the Suez Canal as a short cut to Europe, Asia, and North America.

Page 65

1. a. plateaus b. mountains c. coastal plains
2. a. tropical desert d. tropical steppe
   b. Mediterranean e. varied mountain
   c. Middle latitude steppe
3. a. arid – 0-10 inches of precipitation per year.
   b. semiarid – 10-20 inches of precipitation per year.

Page 66

A. The Arabian Peninsula is the *largest* peninsula in the world and is located in the *Middle East*. The country of *Saudi Arabia* makes up a large part of the Arabian peninsula. It is *865,000* square miles in size and has a population of about *11* million. In the year of *1932*, Saudi Arabia became an independent country. Most of the people of Saudi Arabia follow the *Islamic* religion. They are called *Muslims*.

B.

South	*North*	*East*	*West*
North Yemen	Kuwait	Qatar	Egypt
South Yemen	Jordan	Bahrain	
Oman	Iraq	United Arab Emirates	

Page 67

1. a. narrow coastal plain b. mountains c. plateau
2. Plateau
3. a. Jidda b. Mecca c. Medina d. Riyadh
4. Riyadh
5. a. hot and dry
   b. Nighttime temperatures are cooler.
   c. Cooler months are still warm
   d. Rainfall is scarce during the cooler months, but the grass will grow.
6. Higher lands are always cooler. Coolness produces change in temperature and possible precipitation.

Page 68

1. Paragraphs will vary.
2. a. coffee b. wheat c. dates
3. They provide daily needs of the herders.
4. a. They are very strong. b. They can go for days without water.

Page 69

1. It is rich in oil.
2. a. It is used as fuel for industries.
   b. It is used as gasoline for automobiles.
   c. It is used for heating buildings and homes.
3. Industrial countries depend on oil. High prices produce higher costs of goods and services.
4. If and when oil runs out or the demand drops, the new industries would still provide work, income, and benefits to the country.

Page 70

1. Ankara
2. Damascus
3. Beirut
4. Jerusalem
5. Amman
6. Nicosia
7. Baghdad
8. Tehran
9. Cairo
10. Riyadh

Page 72

1. Over 9,400,000 square miles
2. a. Asia b. Africa
3. a. Arctic Ocean c. Atlantic Ocean
   b. Gulf of Mexico d. Pacific Ocean
4. Two large countries in North America are the *United States* and *Canada*.
5. North America also includes the country of *Mexico* and the countries of *Central America*.
6. a. mountains c. plateaus e. low mountains
   b. plains d. river valleys
7. Mt. McKinley
8. Death Valley
9. Frigid means *cold*; temperate means *moderate*, and tropical means *hot*.
10. All 12.

Page 74

A. 1. B 2. A 3. D 4. C 5. E

B. 1. a. Atlantic Coastal Plain b. Gulf Coastal Plain
2. Appalachian Mountains
3. a. Piedmont Plateau c. Appalachian Mountains
   b. Appalachian Plateau
4. Between the Eastern Highlands and the Rocky Mountains
5. a. Central Plains b. Great Plains
6. It separates the Central Plains from the Great Plains.
7. In the Southwestern part of the Central Plains.
8. a. Rocky Mountains c. Cascade Mountains
   b. Sierra Nevada d. Coastal Mountains
9. The Intermountain Plateau extends from *Washington State* to *New Mexico*.

Page 75

A. 1. Atlantic Coastal Plain
2. Gulf Coastal Plain
3. Eastern Highlands
4. Central Plains
5. Great Plains
6. Ozark Highlands
7. Rocky Mountains
8. Intermountain Plateau
9. Pacific mountains and valleys

B. 1. a, d 2. b, c 3. e 4. f

Page 76

Tropical climates

a. tropical desert b. tropical savanna

Middle latitude climates

a. Mediterranean
b. humid subtropical
c. marine west coast
d. humid continental
e. middle latitude steppe

High latitude climates

a. taiga b. tundra

Page 78

1. In the northeastern part of the United States.
2. Answers should include any three of the facts given on page 77 under "The Humid Continental Region."
3. a. taiga d. middle latitude steppe
   b. tundra e. tropical desert
   c. varied mountain f. tropical savanna
4. Answers will vary.

Page 79

1. a. Mediterranean c. humid continental
   b. humid subtropical d. marine west coast
2. Answers should include any three of the facts given on page 77 under "The Mediterranean Region."
3. In the southeastern part of the United States.
4. Answers should include any three of the facts given on page 77 under "The Humid Subtropical Region."
5. Along the west coast of the United States. Northern California, Oregon, Washington, and western Alaska are in this climatic region.
6. Answers should include any three of the facts given on page 77 under "The Marine West Coast Region."

Page 80

1. a. Northeastern Region c. Southern Region
   b. North Central Region d. Western Region
2. east–Atlantic Ocean south–Gulf of Mexico west–Pacific Ocean

Page 81

	State	*Capital*
1. a.	Maine	Augusta
b.	Vermont	Montpelier
c.	New Hampshire	Concord
d.	Connecticut	Hartford
e.	Massachusetts	Boston
f.	Rhode Island	Providence

2. a. machinery c. shoes e. tools
   b. textiles d. paper products
3. a. tobacco b. maple trees c. hay d. cranberries
4. Students may list any three of the following industries: manufacturing, mining, shipbuilding, fishing, trade.

Page 83

	State	*Capital*
1. a.	New York	Albany
b.	Pennsylvania	Harrisburg
c.	New Jersey	Trenton
d.	Delaware	Dover
e.	Maryland	Annapolis

2. Much of the trade of our country goes through this region.
3. a. wheat b. corn c. fruits d. vegetables
4. Students may choose any four of the following:
   a. raising poultry c. coal mining e. shipbuilding
   b. raising livestock d. manufacturing

Page 84

1. Ohio 3. Indiana 5. Wisconsin 7. Iowa
2. Michigan 4. Illinois 6. Minnesota 8. Missouri

The eastern parts of:

9. Kansas 10. Nebraska 11. South Dakota 12. North Dakota

Page 85

1. Virginia 5. Georgia 9. Louisiana
2. West Virginia 6. Florida 10. Arkansas
3. North Carolina 7. Alabama 11. Tennessee
4. South Carolina 8. Mississippi 12. Kentucky

The eastern parts of:

13. Texas 14. Oklahoma

Page 86

1. The western parts of:
   a. Texas c. Kansas e. North Dakota
   b. Oklahoma d. Nebraska f. South Dakota
   The eastern parts of:
   g. Montana h. Wyoming i. Colorado j. New Mexico
2. Between the Central Plains to the east and the Rocky Mountains to the west.
3. The lack of precipitation causes crop failure and hard times for the farmers.
4. Students may choose any two of the following facts:
   a. A small ranch can be about 1,000 acres.
   b. A large ranch can be about 50,000 to 100,000 acres.
   c. Ranches are large because there is only a small amount of grass for the animals to feed on.

Page 87

1. Parts of the states of:
   a. Washington c. California e. Utah g. Arizona
   b. Oregon d. Idaho f. New Mexico h. Nevada
2. Between the Rocky Mountains to the east and the Sierra Nevada and Cascade Mountains to the west.
3. a. wheat b. sugar beets c. potatoes d. fruits
4. a. mining b. lumbering c. tourism
5. a. Grand Canyon c. Hoover Dam
   b. Bryce Canyon d. Yellowstone National Park

Page 89

1. Parts of the states of:
   a. Washington b. Oregon c. California
2. a. marine west coast b. Mediterranean c. tropical desert
3. The climate is dry. Irrigation provides water for the crops to grow.
4. Students may choose any four of the following:
   a. lumbering - a very big business
   b. fishing - Many fish are caught and sent to canneries.
   c. canning - Fish are sent to the canneries and packed for shipping.
   d. mining - done in California and Washington.
   e. manufacturing - Many important products are made, such as airplanes, ships, electronic equipment, automobiles, machinery, and paper.
   f. motion picture - The center of this industry is California.
   g. tourism - People like to visit the region to see the beautiful scenery.
5. a. San Diego, California d. Portland, Oregon
   b. Los Angeles, California e. Salem, Oregon
   c. San Francisco, California f. Seattle, Washington

Page 90

Alaska is our *largest* state. The population of Alaska is about *534,000*. The United States bought Alaska from *Russia* in *1867*. In *1959*, Alaska became our 49th state. Much of Alaska is very mountainous. The highest peak is *Mt. McKinley*.

The Eskimo people live in the *colder* part of Alaska. Most of the cities are along the southern and *southeastern* coasts of Alaska. Farming is carried on in the *Matanuska* Valley, which is near the city of *Anchorage*. Other important cities of Alaska are *Fairbanks*, *Sitka*, *Ketchikan*, and *Juneau*.

With the new *Alcan* highway and the discovery of *oil*, Alaska can no longer be called the *ice box*.

Page 91

Paragraphs will vary.

Page 92

1. It is the fourth largest.
2. The area is 3,679,245. The population is about 240 million.
3. a. Canada c. Atlantic Ocean
   b. Gulf of Mexico, Mexico d. Pacific Ocean
4. a. Atlantic Coastal Plain b. Gulf Coastal Plain
5. a. Central Plains b. Great Plains
6. a. Piedmont Plateau c. Appalachian Plateau
   b. Appalachian Mountains
7. a. Ozark Plateau b. Ouachita Mountains
8. a. Sierra Nevada c. Pacific Coastal Ranges
   b. Cascade Mountains d. Rocky Mountains
9. 10
10. a. Mediterranean c. marine west coast
    b. humid subtropical d. humid continental
11. a. New England states e. Great Plains states
    b. Middle Atlantic states f. Intermountain states
    c. North Central Region g. Pacific Coastal states
    d. Southern Region
12. 1959